MICHAEL JAEYOUNG L

MyData

and

Data Sovereignty

in the

Age of AI

Including Case Studies in Korea

poppypub

MyData and Data Sovereignty in the Age of AI by Michael Jaeyoung Lee

Published by POPPYPUB in Hoboken, NJ
www.poppypub.com
poppypub is a trademark of POPPYPUB LLC.

ISBN 978-1-952787-29-4 (paperback)
ISBN 978-1-952787-30-0 (ebook)

CONTENTS

Preface

Are there other concepts in the data industry that are as new and frequently mentioned as data sovereignty and MyData? It proves that various innovative attempts are being made in the industry. However, the use of personal data is likely to conflict with the value of privacy protection. Gaining insight into users through big data or machine learning is possible. However, to apply it in reverse, there is no choice but to identify and target the individuals profiled through the insight.

This book introduces and proposes new approaches to safely utilize personal data for companies and institutions that plan or operate MyData services based on personal data sovereignty. In particular, convergence services that provide new user value in combination with AI technology are expected to emerge. There are also voices of concern that the deregulation to revitalize the MyData industry could eventually negatively affect security and privacy. These days, when trying to find new added value in data combinations, the personal and social impact is inevitably significant in the event of a privacy accident. This is because the more combined data, the higher the sensitivity. Therefore, this book even proposes a new approach to solving the privacy problems that may arise from using personal data. I tried to deal with MyData and data sovereignty in the AI era, but I also tried to present case studies in Korea.

This is my fifth book and the first book I wrote in English, which is not my mother tongue. I had many difficulties selecting words and making proper sentences. I ask the readers to be generous if they encounter slightly awkward words and sentences. Please get in touch with me directly if you need help understanding or have any misunderstandings. While writing, there were many ups and downs at work and at home. There were so many dramatic episodes that seemed to be much more fun and would sell better if I wrote them as a book—I'll do it later anyway—I am always grateful to my colleagues and juniors who supported me when I was thinking about the new challenge of starting a startup company. I was also worried about whether I could finish writing while running my company, but I could eventually put it out to the world.

I express my infinite gratitude to my family, friends, and colleagues for always supporting me silently.

I.

The Advent of Data Sovereignty Era and Data-Driven Society

Background

We are in the data era. We live in the information age, and data affects every part of our lives. As the years go by, our lives are increasingly defined by data generated from all interactions, including the Internet of Things, browsing, automobiles, e-mail, and hospitals. Data insights also affect business strategies, customer service, and national policies. Data sovereignty is sometimes used to mean that it is subordinate to the law and governance of the country, where data is collected from the perspective of sovereignty between nations. In particular, China tends to put the country's data rights ahead of individual data rights. The Cybersecurity Law of the People's Republic of China enforces the storage of data generated in China. It stipulates the technical cooperation required by the government as a duty of companies. There are even mandatory regulations that require companies and institutions to provide information for data decryption if required by the government. However, when it comes to data sovereignty, it is mainly limited to personal data. It is used in the sense of the right to decide where, how, and for what purpose one's data will be used by granting rights to individuals, such as body or property rights.

As data becomes more important, it is becoming more important to use it safely and competently. Misuse of data may result in the opposite of our expectations. Although humanity has been civilizing and developing culture and industry through fire, misusing fire can instantly turn everything achieved into ashes. Using data reasonably and safely for good purposes can make our lives more convenient and prosperous. However, it can also lead to irreparable consequences, such as privacy infringement caused by personal data leakage. Therefore, to revitalize the data economy, the safe use of data must come first. Even in the so-called the 4th industrial revolution in which all industries and information and communication technology converge, data is ultimately the foundation of everything. Strategies related to the 4ᵗʰ industrial revolution by country, such as the U.S.' 'Industrial Internet,' Germany's 'Industrie 4.0', China's 'Made in China 2025', and Japan's 'New Industrial Structure Vision,' are focused on advancing a data-driven society by maximizing data competitiveness.

The Era When Data Determines the Course of Life

Data is deeply embedded in every corner of our lives. However, in most cases, individuals do not know where their data is stored. Not only information directly entered by individuals and data generated through service use, but also data that is unknowingly collected through numerous sensors is more difficult for individuals to notice. Globally, platform companies such as Facebook, Amazon, Netflix, and Google, called FANG, hold the most personal data of individuals. But that's not all. Sensitive data, such as financial or medical data, are stored in different institutions such as banks and hospitals.

In addition to sensitive data, various types of data are generated daily. This data is stored and managed by companies and institutions that provide services. Companies use this data for analysis to gain insight into individuals and for targeted marketing to provide services tailored to individual interests and preferences. Increasingly, we are becoming a data-driven society where the actual world and data are connected in sophisticated ways.

In this society, our data will determine the course of our lives, even if we are unaware of it. For example, you can rely on restaurant recommendations from search engines for places to eat. Similar behavioral patterns are observed when shopping online. Perhaps we are already accustomed to a data-driven society. In addition, as COVID-19 requires non-face-to-face contact in all areas of society, entry into a data-based community is accelerating. As we enter the post-COVID era, there is a joke that COVID-19, not corporate CEOs, CTOs, and CIOs, accelerated companies' digital transformation in the traditional industry sector.

However, in this data-based society, the consideration of the position and role of an individual is insufficient. While the quality and speed of decision-making can improve as our lives and data become increasingly sophisticated, we as individuals must be able to use data intelligently to benefit the information subject. The starting point of discussion about such an environment is bound to be sovereign. Ultimately, data holders can significantly influence the lives of individuals indicated by data than individuals expect. Hence, discussions about data ownership become a necessity, not an option. In a data-driven society that can build sophisticated links between data and users' actual activities, there is growing concern about who holds what data, whether it starts to affect our daily lives directly, and whether individuals can fully understand it. If medical data becomes the standard in determining how to provide medical care, it may become a system where treatment cannot be obtained without medical data. Suppose a system is introduced that determines which cars can be driven based on driving record data. In that case, it may become a society where purchasing a car without a driving record is impossible. Not being able to provide financial products and services without proper financial data is not new and has already become a social problem, so several attempts are being made to provide finance for the so-called 'Thin Filer.' Our society is already too dependent on data to dismiss the situation in which data holders unintentionally influence each individual as imaginary. The problem is that the personal data that defines the individual is still in others' hands. There are many cases where I don't even have a problem with others having my data because I've gotten used to it for so long. To prevent this, in the end, individuals must have the right to control their personal data, and the right to my data results in data sovereignty. It will be a process of drawing a social consensus to restore data sovereignty.

A Data-Driven Society: Capturing the Value of Data

We often don't realize how much data we are generating. We are generating data at the business level and also regarding individuals. All this data will open up a world of unique user experiences and new business opportunities. The first step in creating data value is, of course, securing and collecting data. IDC (International Data Corporation) predicts that by 2025, nearly 20% of the world's data will be critical to everyday life, and about 10 % of that data will be critically important. For example, if a company wants to study how its product affects a particular market segment, you will find the relevant data first. However, if it is not correctly processed or is missing, an analysis of the data will yield information of much less value. Therefore, better tools are needed to support data visualization, data analysis, and data collection methods. This is why new technologies, including machine learning and data warehousing, have emerged.

Through the Data Age 2025 white paper, IDC expects the amount of global data to be analyzed to increase 50 times to 5.2ZB in 2025 and the amount of analysis data covered to increase by about 100 times to 1.4ZB in 2025. Of course, the value will increase as much as the increased amount of data, and who takes the increased value will result in a question of power in a data-based society. Capturing this value will eventually be the first entity that owns the data and can analyze it. So, by creating various logic and reasons, groups such as specific companies and institutions are trying to collect and hold data. For example, in the case of Japan, it is said that an information bank is necessary to protect the rights of individual consumers. Although the function of the information bank is at the direction of individual consumers, the argument is that information banks will defend the rights of individual consumers as part of a social infrastructure system.

Of course, these logic or arguments cannot be criticized as absurd or undesirable. Japan's information banks differ partly from institutions currently defined as PDS (Personal Data Stores; this will be covered later) and are designed to support data distribution according to individual consumers' instructions. However, it will take time to determine whether information banks can prioritize public interest purposes for the better lives of people instead of empowering digital platform companies and other companies.

When discussing a data-based society, an approach is usually taken from a national or industrial perspective. For example, discussions about the need to strengthen national statistics governance in line with the data age or the need to pay attention to enhancing new dominance based on data due to the birth of Big Brother, including data dinosaur companies, are examples. Therefore, if our society evolves into a data-based society, social consensus must be premised on all matters. However, social discussion on this has yet to begin. At most, the degree to which the non-face-to-face economy was revitalized with COVID-19 and that local restaurants started to provide non-face-to-face services through delivery apps actively explains the phenomenon that data-based services are rapidly spreading in people's daily lives. There are still many topics in the gray area, starting with the issue of data ownership, the right to use it, and the scope to use personal data for the public good. Many agendas will emerge after the advent of a data-based society.

Data: The Key to Business Survival in All Industry

Beyond certain behaviors or predictions for the foreseeable future, data is now a matter of company survival. The enterprise's sustainability cannot be guaranteed without clear and comprehensive insight into the data. Data-driven insights are becoming the biggest differentiator for business in many industries. Whether organizations attract new prospects, serve existing customers, or protect assets, data utilization strategies are now essential. Many say that data science is the core of the 4th industrial revolution. Still, in practical and realistic terms, it has become a world where companies can promise survival only if they can properly and effectively analyze data.

It can analyze regularity and insight from historical data to estimate future markets or risks, discover new markets, and provide personalized services by analyzing unconscious behavior patterns of existing customers or use them as corporate management strategies such as cost reduction. Big data has been expanded to include unstructured data rather than just large amounts. It has become a concept that includes analysis technology that extracts new values and derives results from it. Big data is now gaining importance across all industries, including politics, economy, society, culture, and science and technology, and competition without borders is fierce in the data industry, not only in advanced countries in the United States and Europe but also around the world. Developed countries have already been accustomed to purchasing and using various data types for multiple purposes, such as exploring new markets, developing new products and services, and improving corporate efficiency since the 1970s, long before Clive Humby said data was new crude oil in 2006. To overcome those familiar cases, social preparations are needed to support them and the legal and institutional environment.

It has become essential to effectively analyze unstructured big data and well-organized data in the database to use big data well. In particular, it has become crucial to effectively analyze big data such as social network services, the Internet of Things (IoT), and sensor data. For big data analysis, technologies such as Hadoop and NoSQL, various analysis methods such as statistical processing and data mining, and artificial intelligence technologies such as machine learning are used.

It has been over ten years since everyone shouted big data in one voice, and it is clear that big data is a crucial factor in leading the post-COVID-19 era, which is opening up a non-face-to-face society. However, limitations in using big data have been exposed in recent years regardless of field. Behind many advantages, such as providing customized services and increasing corporate profitability, are side effects, such as privacy issues. A case in point is a social controversy in 2018 when Cambridge Analytica was found to have collected profiles of millions of Facebook subscribers without personal consent and used them for political propaganda. Even if it is not a leak accident, collecting individual data eventually creates big data. Of course, companies and organizations that observe compliance will go through a de-identification process. However, re-identification issues still arise when these small data are gathered and become a large data set. Collecting as much data as possible is necessary to increase the utilization of big data, so a trade-off is inevitable. Finding a balance point that can utilize big data while protecting personal data privacy is essential. However, compromising is sometimes difficult since privacy is based on human rights.

EU Data Protection Legislation and Emergence of MyData

Even before the GDPR (General Data Protection Regulation, the European Union's data protection law) was enacted, Individual European countries had laws to protect their personal data. However, while the increase in cross-border trade allowed for the free international flow of information needed to support international trade, new standards were also required for individuals to control their personal data. The problem was to create a balance between national concerns about individual freedom andprivacy and supporting free trade from the perspective of the EEC (European Economic Community).

In many cases, legal requirements related to personal data protection were often misunderstood as unexpected regulations. However, the right to privacy is considered a fundamental human right. The Human Rights Declaration, adopted by the United Nations General Assembly on December 10, 1948, was a clear starting point for setting standards for protecting individuals.

Article 12 of the Declaration of Human Rights clarified that privacy is part of human rights by declaring, "No one shall be subjected to arbitrary interference with his privacy, family, home or correspondence, nor to attacks upon his honor and reputation." In 1950, the European Council asked individual European countries to sign the ECHR (European Convention on Human Rights), an international treaty to protect human rights and fundamental freedoms, and it went into effect on September 3, 1953. Since then, from the late 1960s to the 1980s, several countries, including Europe, have implemented legislation to control the use of personal data by government agencies and large companies. In the early 1980s, the OECD (Organization for Economic Cooperation and Development) created the OECD guidelines on privacy and protection of cross-border personal data flows, forming basic rules that control cross-border data flows and enable harmonization of cross-border data protection laws. The OECD has made efforts to maintain consistency with the principles developed on behalf of the European Council, which is not much different from the Convention 108 related to automatic processing of personal data later adopted by the European Council.

The European Commission submitted a proposal to the European Parliament in early 1976 to develop guidelines to harmonize data protection laws, which was implemented as Directive 95/46/EC (Data Protection Directive). In 2009, a comprehensive review of the legal framework for data protection began, and in 2010, a strategy was established to strengthen data protection rules. As a result, in January 2012, a comprehensive reform of the guidelines was proposed to the European Commission in the form of a law that applies a set of rules throughout the EU, which was the draft GDPR.

The GDPR is designed to create a "strong and more consistent data protection framework" backed by strong and consistent enforcement to "build trust that enables the digital economy to develop in the internal market." In addition to the GDPR, the EU is completing the GDPR framework. In addition to the GDPR, the EU is completing the framework for GDPR through related legislation such as the LDPD (Law Enforcement Data Protection Directive), the ePrivacy Directive, etc.

It is clear that data is a resource like oil for companies

that use it, but it has been useless to the individual who provides it. As a result, individuals refrain from providing personal data as much as possible, and companies struggle to use it in a valuable way through data collection, combination, processing, and analysis due to various compliances. Various attempts are being made to solve this problem, and MyData is the most noteworthy among them.

MyData does not simply mean 'my data.' It means that a series of legal, technical, and administrative measures are taken by an individual who is the data subject of personal data to manage, control, and utilize their information so that the individual has complete decision-making power over personal data. It aims to be a human-centered model for managing and using personal data. It gives individuals the right to self-determination over personal data, allowing individuals to directly give companies and institutions the right to use personal data such as finance, shopping, health, or health data.

Personal data is rarely limited to the individual but instead generated through interaction. Even the date of birth is data about someone who gave birth to me, and my click is also part of that posted web page. Individuals inevitably exist in a meaningful form within groups, communities, and societies that are inseparably connected. As a process and result of interaction, MyData is being established worldwide. The MyData Global Association, an international non-profit organization established through a series of international conferences and discussions, aims to empower individuals by improving their self-determination of personal data.

Limitations of MyData
and the Process of Overcoming Them

However, MyData is not the solution to everything. MyData, which restores sovereignty as the owner of my data, is also taking various forms of access. At the end of this book, I would like to present a new MyData approach regarding data utilization governance. In particular, most of the existing legal systems and technology stacks are premised on data processing by companies and institutions. It is necessary to seek deep and new introspection from various aspects and perspectives. For example, to discuss MyData, in the end, it is essential to discuss personal data rights, and academic, legal, and institutional considerations are needed.

Data is a fundamental resource of the future society to the extent that the word "data is the new oil" is now commonplace. Discussion on technologies and platforms that can provide individual data sovereignty in an environment where a small number of data giants monopolize, collect, and utilize vast amounts of data is a necessary agenda regarding the data industry and political, economic, and social aspects. However, it is also essential to consider the risks of introducing MyData. Personal data transmission, collection, and analysis are smooth and ostensibly under human control. However, acknowledging broad consent to using personal data leaves individuals unable to make real choices, like in the movie 'Matrix.' In the early days of the Internet service, consent to collect personal data was required. Still, in the end, it was not an individual's actual choice, so in some countries, it was divided into "essential," "functional," and "marketing" to obtain consent. In the meantime, some companies have written small and too detailed to induce consent without reading, but in the end, it isn't easy to recognize such consent as proper legal consent. Unless you give up your data, whether you like it or not, the tension between those who want to use data for free and those who do not wish their life to be interfered with is a complex problem to resolve. MyData is expected to bring long-awaited digital rights and power to people, but at the same time, individual responsibilities increase. First of all, individuals should manage the location of personal data storage and usage. It will eventually be abused if you don't manage it because the ability and opportunity to manage individual data problems are unevenly distributed. Some people are good at managing, and others are not. More details about MyData will be covered later in this book.

Personal Data Sovereignty

In the past, the term consumer data sovereignty has often been used when discussing sovereignty over personal data. Along with the term data economy, it was used to refer to the rights of consumers, who are data subjects, as the awareness that consumers' rights and interests should be protected has arisen. However, because of the passive meaning given by the word 'consumer' itself, it is not a suitable term for actual personal data self-determination. The term consumer data sovereignty partially reflects the industry's position. In fact, while referring to consumer data sovereignty, the logic of claiming that it is for the development of the data economy is based on a business-friendly view. In accordance with the privacy acts in countries, shifting from a data utilization-oriented policy perspective to a protection-oriented regulation and security-oriented perspective is necessary. First, it is essential to seek ways to secure practical transparency in data processing, away from the prior consent-oriented system for personal data collection. A certain level of data processing is essential to provide goods and services that match the data economy, and regulating it only with the existing prior consent system may limit the smooth use of services and increase consent fatigue, hindering consumers' perception of privacy. It may be more effective to ensure transparency in the data processing process and to guarantee the right to follow-up data processing based on this rather than the prior consent procedure. The economic value of consumer's personal data should be clearly calculated and allowed to be selected by consumers so that consumers do not belong to specific goods or services. It is also necessary to discuss ways to realize consumers' active interests as data producers, considering the role of a confident economic entity that creates added value through consumer data generation.

As of 2023, the serviceable MyData model has many limitations regarding the above perspective. But this limitation will be overcome in another form, and the clue seems to start with the awake and prepared individual eventually. When the Internet era opened, individuals had nothing. To receive the service, an individual starts by creating an account, which includes identifier information and credentials needed to log in. It will not be easy to improve the consciousness of individuals because they have become so accustomed to it. However, individuals are now accustomed to high-tech devices and are no longer underdogs in the IT world. Ultimately, based on IT, individuals will try to become the center of the ecosystem to restore their rights. Recently, many studies and demonstrations have been conducted on DID (Decentralized Identification), which is introduced through decentralized ID and distributed identity authentication. DID is also one of the suitable mechanisms to utilize personal data while protecting privacy. However, it is still not a completely decentralized model as it is a form of identity verification based on the premise of a separate identity issuing authority. Ultimately, a social consensus on my proving myself without separate proof from an issuing authority will likely be the last step in establishing individual data sovereignty. However, it will be a distant future.

Concerns About a Data-Driven Society

Unlike the digital world in the early days, the activation of mobile technology and social media has enabled people to produce and release various information in real-time. As a result, the era of big data, where much data is accumulated, has opened. When there is much data, new information appears based on this. A new era of knowledge revolution is opening as such information is processed through artificial intelligence. In the future, the amount of data will undoubtedly increase faster than in the past. Smartphones generate a lot of data from time to time, including voice and text, as well as photos and GPS location information. Social networking services are pouring out messages from countless people worldwide in almost real-time. In the future, various actions in the mobile environment will all become new data and be transmitted to some repository in cyberspace. Smart technology will also be applied to automobiles, home appliances, and buildings, becoming a source of countless data. People can access this data anytime, anywhere, through mobile devices such as smartphones. Above all, properly utilizing and processing big data will be a significant competitive edge shortly as services that collect and use such generated data are activated. The actual value of data is determined by whether a new social value can be extracted from the data. Obtaining a sufficient amount of data is only in the early stages. The next step is to manage and analyze the data collected like this to give answers on how to extract real business and social value. Through this stage, data will be serviced and distributed in various ways, and it will evolve into a stage that creates hidden values that have not been thought of in the past. After all, from this point on, our society is data-driven.

Big data enthusiasts highly value the possibility of analyzing accumulated data and solving many problems in our society and business. They refer to various scenarios such as making correct judgments using data, enjoying a healthier life, making effective urban planning, planning better-selling products, and creating a safer society. There is also a slightly extreme case, which is the case with the U.S. National Security Agency. They believe that terrorism can be prevented by collecting and analyzing data from all call records. The intentions are understandable, but they can come very horrible to someone. Side effects that may arise in the era of big data should be considered. The first side effect that comes to mind is the issue of privacy. Data is eventually created by collecting information from numerous devices or individuals.

We must gather as much personal information as possible to analyze and benefit from this data. Each individual must also be able to use a lot of personal information to obtain information that suits them through data analysis. This means that to benefit from data as much as possible, as many people as possible should agree to be used by others rather than protecting personal information. This is like a double-edged sword. In other words, personal information protection cannot be claimed too much to receive data-based intelligent services. On the contrary, to strongly protect personal information, one should not give one's data as much as possible, but one should also be prepared to be alienated from obtaining necessary information and knowledge. In the end, efforts to find a balance between the use of data and personal privacy protection at a socially valuable, ethical level and with no significant legal problems are substantial.

The era of big data can have a side effect of loss of personality. Imagine that there are various forms of 'recommendation' by gathering data from many people. Each individual's decision is likely affected by what is recommended through data analysis. Take Facebook News Feed as an example. News Feed analyzes individuals' "likes" and "comments" to prevent excessive information from bothering users by showing or not showing appropriate content. This seems convenient, but when you think about it, the system's decision is blocking my opportunity to see the information itself. Only those that fit the user's disposition, such as purchasing goods, choosing services, and opinions, are exposed, and if this continues to be repeated, the situation will become even more serious. This is because even the opportunities to try the experiences I hated or was reluctant to do in the past may disappear. Already, many people only look at the information that search engines choose and consume only products that have high ratings or are customized. In this case, an individual's choice, or 'individuality,' is challenged. This is why the situation in which people are accustomed to making easy decisions as they are "recommended" and are castigated for their unique choice skills does not sound just like stories in science fiction.

Finally, there are concerns about the birth of big brothers, such as giants and governments, and the strengthening of their control over citizens. Companies and public institutions that provide data services say not to worry, but it is unbelievable, considering that data can also be a power.

II.

Ownership of Data
(Big Data and Personal Data)

Background

There are few cases where there is a socially explicit agreement on who has the right to records and data about individuals created through services. It is still a constant debate among policymakers, economists, and consumer activists. As the utilization of personal data is increasing, clear definitions and social agreements are required, whether it is ownership or right of use. Spotify, the world's largest Internet music streaming app, was born in Sweden in 2006 and is famous for its "service that knows my taste better than my old lover" online. The "Recommendation for You" service is popular and recommends new songs weekly based on data from 250 million users. Like Netflix, it is a data giant that collects data to target users' tastes. Free Spotify users often think that ads are the price of free streaming. Advertisements account for only 10% of total sales. Free users provide preference data without their knowledge. Spotify was able to increase its share around the world thanks to the heart buttons pressed by free users. It's not just Spotify. For tech giants, data is money. Each individual's actions recorded one by one create new added value. However, technology giants do not benefit individuals who generate such data financially. Meanwhile, while industrial field jobs are decreasing and universal labor faces a crisis due to technological advances in AI, threatened individuals eagerly provide their data to technology giants. For this reason, a completely different level of discussion on the 'data economy' is being raised. It is the debate over 'ownership of individuals'.

Discussions about data ownership have existed in academia, but the main character who recently publicized the issue is American businessman and politician Andrew Yang. Andrew Yang put forward a policy to provide a basic monthly income of $1,000 per person. He said he would raise funds for this through "Tech Check." It means collecting a data usage fee from companies. In other words, Andrew Yang argues that technology giants are making money using personal data. However, users who provide personal data do not receive proper compensation. And he claims to redeem it by giving universal basic income. It was Gyeonggi Province in Korea that began the actual data dividend. However, it was the United States that presented this discussion in earnest in the political world. This served as an opportunity to raise the alarm for technology giants hurrying to secure data indiscriminately. Until now, technology giants have received data in exchange for Internet services such as e-mail, search, and SNS. Data that produces content and engages in creative activities on the platform they created eventually requires active action by users. Every act of posting articles, photos, videos, and expressing emotions through posts is a good material for machine learning for tech giants.

Some interpret such individual data production from the perspective of labor. In their book Radical Markets, Professor Eric Pozner of the University of Chicago and senior Microsoft researcher Glenn Weil describe the data giants and data-producing individuals in the labor market. The tech giant is using its monopoly on demand to keep the wages of those who supply data tied to zero, arguing that now could be the time for data workers worldwide to unite and jump into the data labor movement. It is pointed out that it is difficult for individuals to negotiate data provision against technology giants alone, so it is necessary to consider solidarity between data workers. It is in line with the MyData movement.

In Korea, a bill has been proposed to include data,

including personal data, in the definition of 'object' under civil law. The content lays a conceptual foundation so that data can be legally traded. This aligns with the discussion on basic income in that it is a kind of technological development and labor reduction. The view is that data, including personal data, is not a natural resource that can be collected recklessly but "something that has to be paid for." While basic income is based on the government's fiscal execution, these civil law amendments are more of a marketist approach that recognizes individual ownership but allows companies to make transactions at a cost. Although it was not processed within the session, it became a concept of paying individuals based on ownership rather than the concept of data "dividends," which could be viewed as a suitable approach to the capitalist concept, so it could be evaluated as an approach that does not conflict with ideology.

Basic Rights on Data

Among the rights to data, the first thing that comes to mind is the question of 'who owns the data.' For example, when patients' weight is measured and recorded in hospitals, this is only one data. However, the average or variance that statistically corresponds to information and the owner of this information can be said to be in the hospital that performed the calculation. However, depending on the viewpoint, the data itself may mean information. For example, the individual weight data corresponds to critical personal data, so it is controversial whether the hospital owns the weight data just because the hospital collected it with their equipment. To jump a bit, we'd be hard-pressed to agree if I claimed my footprint belonged to Nike for wearing Nike.

The data economy is a term not left out when discussing the 4th Industrial Revolution. The term data economy was first used in the European Union. The EU Commission defines the data economy as an ecosystem created by members who handle data and comprehensively explain the creation, collection, storage, processing, distribution, and delivery of data. In this process, the data economy ecosystem is difficult to discuss except for concepts such as money and value. So, the discussion eventually develops into data as property, that is, the property rights of data and, ultimately, the discussion of ownership. In particular, it is necessary to approach large companies and platform companies, which are likened to data dinosaurs, regarding redistributing capital power in the data economy. The production and distribution of data are often interpreted as shared resources rather than the characteristics of ordinary property rights. And when specific expertise is required to produce data, it may be interpreted completely differently. For example, if a doctor generates a medical record, the right to medical data may be approached differently. The right to data has already been approached by attempting to grant specific permissions to objects or objects digitized in terms of digital files and digital assets. In recent years, such attempts have been made to asset records recorded in the blockchain. Another notable change is that attempts to secure rights to data have continued steadily, and new attempts, such as NFT (non-fungible token), are also ongoing.

Nevertheless, considering the changes that the data-based society will bring, it is necessary to approach it from the perspective of universal human rights. In exchange for using the platforms of data giants, we are handing over almost all of our personal data and unknowingly becoming dependent on various platforms. Instead of providing control of data, we are enjoying convenience. However, since this is not the ideal that humanity should ultimately pursue, the discussion of rights to data will continue.

Not many deny that data is a resource for the digital economy. It is also pointed out that platform companies enjoy most of the benefits from the data, mainly in British and American countries. Tim Dunlop, an Australian political philosopher and author of Future of Everything, argues that profits from data should be linked to basic income. The public should own data because it is created by everyone. The value of individual data is not as significant as the sum of data. In this regard, it is reasonable for the state to come forward and redistribute it as basic income rather than for users to get their share through individual negotiations. Tim Dunlop argues that data is a common good shared by many people, such as underground resources such as oil and iron ore. It is argued that individual data held by a person is not of great value, but the total value of the data is enormous, so many people commonly share it. The data can be seen as similar to underground resources such as oil and iron ore. Companies pay royalties when digging out underground resources because the buried underground resources do not belong to them. Currently, data ownership is not clear, and a small number of entrepreneurs are using it to earn enormous wealth. Of course, innovative companies using data to gain wealth should be encouraged. This is because there is also an aspect of collecting them and creating something in a way that is useful to humanity as a whole. However, the problem is that the nature of the public property of data is not considered at all, and all of them go to specific companies.

Ownership of Data

Surprisingly, no country has yet recognized the right to data in the form of ownership, even in countries where data is stipulated in their constitutions. It is widely believed that most legal interpretations of personal data ownership do not have a legal theory to justify exclusive property rights. Among legal experts, the prevailing opinion is that we should not talk about ownership but focus on data usage rights and some licensing concepts. Instead, it is argued that, unlike general data, discussions on data ownership of individuals who are producers and providers of personal data need to be reviewed regarding fundamental rights at the constitutional level, which is the basis of the law. In some countries, basic rights to data are not specified, but the right to self-determination of personal data has been recognized through the same interpretation as precedents. However, the basic right to information at the constitutional level has already been discussed in some countries, and some are currently in progress.

Some argue that data is subject to ownership due to the nature of private property, but if data is treated only as private property, individuals must have individual relationships with various platforms that use data. Then, they have no choice but to go to a situation where they negotiate personal compensation for their data. The value of individual data is low and only increases when gathered. Therefore, even if some data corresponds to private property, the view that it is desirable to treat data as public goods is still valid. There have also been discussions of data ownership legislation in the EU, but by recognizing personal ownership of data, the obligation is limited to privacy protection in most cases. It is not an active form that gives rewards for data. It just states that it is not permitted to sell data elsewhere without the permission of the individual who gave it. Therefore, discussions often result in dividends. In fact, the U.S. state of Alaska has been operating the "Alaska Permanent Fund" since 1976, in which oil is recognized as a public good and profits from oil sales are distributed to citizens as basic income rather than monopolized by companies. The same basic income distribution can be applied to data, just like the saying that data is new oil. It may be challenging in reality, but the question is whether social consensus eventually comes. In the data-based society just around the corner, existing data-related laws and systems may need to be reviewed from scratch. The concept of basic rights to personal data means that discussions can begin with confirming that the right to control such personal data ultimately lies with the individual. The digital world is changing rapidly based on data, but we, who exist as data within it, have never discussed how discussions of data rights can affect individual fundamental rights. It is also because it is not an agenda that can be approached simply by dividing it into legal logic or technology and industry levels. However, as the discussion develops, personal data reveals problems caused by thoughts and logic that deny the concept of personal ownership, and ultimately, the emergence of new technologies and platforms will change the perception of data ownership.

The legal system of data attribution, protection, and transaction has not yet matured. Therefore, it is premature to expect comprehensive and systematic data legislation. Most problems can be solved through technology, markets, and contracts. Even if data ownership is not recognized, discussing whether the right to redeem individuals' data stored on the cloud server is recognized and whether it can be regarded as the object of trust will be necessary. Just as it took time for a new normative system for intangible objects such as copyright law and patent law to take place, matters related to data ownership will also be organized through this geographical process. According to the "Building a European Data Economy" white paper released by the EU Commission on January 10, 2017, Europe is not properly utilizing the potential of data and proposes ownership of data in the context of the European data economy. Data itself has value as an asset, but the value as an asset is not clearly protected legally now. Therefore, even if it is not an exclusive right, it seems more desirable to be discussed in the form of ownership, which is expected to be approached differently from civil laws in most countries. Numerous jurists continue their research on this, and social consensus is also an issue that must be resolved.

The legal community of each country has raised several questions about the scope of rights that individuals should have. Copyright or other intellectual property rights, trade secrets, or new property types were also discussed extensively. In fact, actual ownership and control is a critical missing piece in recognizing personal data as an entity that individuals can lawfully own.

Each country's private property system is indeed different, and each country has different requirements to be regarded as a "thing" under property law. For example, British common law requires the following four major requirements to be met to regard the object as an 'object' from a strict property law perspective, according to Mummery LJ's opinion on Fairstar Heavy Transport NV v Adkins [2013] EWCA Civ 886 (19 Julie 2013). The first is certainty, the second is exclusivity, the third is control, and the fourth is assignability.

Data stored on an individual's device (especially a personally owned smartphone) can be considered to meet all such requirements. Individuals can be positively identified (certainty) and block other users from accessing their data (exclusiveness). In addition, individuals can exercise control over that data (control), and You may assign or transfer such data to third parties if you wish. In this environment, it seems that it is possible for social consent at the level of common sense that individuals can 'own' their personal data. In fact, this logic has no room for refutation by the utilization entities in that it is the same as the argument that companies make about the data they collect. Suppose it can be argued that data collected over time forms a proprietary asset and trade secret. In that case, individuals should also be able to assert ownership of the data they store. More remarkably, the argument that an individual is the rightful owner of personal data stored on an individual's device does not require any change from the existing legal framework. Now, solutions that solve the problem of data ownership by actively utilizing IT devices such as smartphones can be the beginning of solving the problem of personal data ownership.

Critical Views on Data Ownership Theory

Nevertheless, the concept of data can be fluid. The value of data depends significantly on the context of its utilization and whether it is combined with other data. One common question is how much data costs. In my case, I answer that the data is valuable, but there is no fixed price. The relationship of rights to data is also flexible. Since ownership is a concept that has a long legal system as a strong property right, it isn't easy to apply ownership to data. Most lawyers say it is challenging to be recognized under the current law. It is also said that establishing a law will take a long time. This is also the case with the history of establishing intellectual property laws. However, data use and protection may not be achieved only through law. Unfortunately, however, legal protection by control, represented by ownership, has initially been centered on fluids. These rights were formed a long time later than ownership of fluids. The copyright was first introduced in 1710 through the Statute of Anne in England, and it was not until the 19th century that it became a legal right. Personal rights are the product of relatively modern law, and the establishment and expansion of personal data-related rights systems have taken place at a remarkable pace in recent decades. There were two triggers here. The first is that privacy has shifted from a domain concept to an information concept, and the second is that 'information' has conceptually combined with 'self-determination.' The concept of self-determination that humans decide their own destiny has been used mainly in connection with important legal interests such as life, body, and fidelity. However, as personal data conceptually combined with the right to self-determination, personal data has become one of the important legal interests. However, whether it falls under personal data here is easy to expand because it is based on the 'possibility' that can be identified. Based on this, the scope of personal data protection has continuously expanded as the personal data protection legislation has been comprehensive and systematic.

An aspect that complicates the discussion is the expression 'ownership.' Ownership has slightly different meanings depending on the country. In German law, complete exclusive control over fluids is called ownership, and in France, exclusive rights to intangible objects such as bonds are also included in ownership. In Anglo-American law, ownership is the exclusive right to occupy and possess something. Whether ownership of data can be conceived depends on whether exclusive control over data in the relevant legal order can be called ownership. In Korea, the penetration rate of electronic medical records is high, and the health insurance information of the entire nation is secured, so the healthcare data industry has perfect conditions for development. Under these circumstances, some arguments acknowledging ownership of healthcare data to healthcare workers will play an essential role in the development of the data industry and that patients should be able to recognize data ownership to individual patients and sell their data to healthcare workers so that they can participate in the economic benefits created by data utilization. Healthcare data also considers the purpose of a separate law on health care, but the conclusion does not change from the general theory of data ownership. Health care data corresponds to sensitive information likely to infringe on the data subject's privacy significantly and consists of medical and health information and the patient's right to self-determination of personal data, personal rights, confidentiality, and freedom of privacy. In particular, health care data is insufficient to protect through the right to self-determination of personal data, such as prior consent. This is because patients tend to give up their right to self-determination of personal data for the more important benefit of treatment purposes. Since the close trust relationship between doctors and patients makes the benefits inherent in the right to self-determination of personal data meaningless, more care should be taken in protecting health care data. This situation applies equally when data from heterogeneous industries are combined. This is because combined data is likely to be sensitive data.

Big Data and Personal Data

Companies use personal data almost free of charge and turn it into an asset. Corporations invest heavily in personal data processing for their benefit, but individuals clearly have rights to their personal data. Sometimes, personal and corporate goals for personal data don't align, and that's where a conflict of values arises. Big data can be defined in various ways according to use cases, but to define big data narrowly, it can be defined as large-scale data containing text and video data as well as numerical data. According to the McKinsey report, "Big data" refers to datasets whose size is beyond the ability of typical database software tools to capture, store, manage, and analyze. Of course, in the past, large amounts of data were collected, processed, and utilized. However, with the development of IT, a large amount of data is now being created that is difficult to measure. In addition, with the development of high-speed networks, data transmission and reception time and cost have decreased, and data storage, processing, and analysis technologies have also developed rapidly, changing differently from the past. The same is true in terms of use. Beyond simple analysis, it began to be used to detect risks or predict the near future by gaining insights through various pattern extractions. Now, even in sports games, it has become a natural manager's role to analyze data on the opposing team to extract patterns and devise a winning strategy.

The concept of personal data is central to data protection law, and its definition is deliberately broad. Since there is a "Personal Information Protection Act" in Korea, there is a tendency to prefer the term 'personal data' over 'personal information,' but there is no difference between personal information and personal data in practice, and it is okay to use them interchangeably. Even long ago, data and information were distinguished by defining data as information when processed. However, since untreated data is challenging to have meaning in itself, it tends not to differentiate between data and information, particularly now, even in areas related to personal data. In the U.S., we use two kinds of terms, 'Personal Information' and 'Personally Identifiable Information (PII)'. We can regard personal information as almost the same as personal data. Those were used differently depending on jurisdiction. Simply put, we use personal data in GDPR, and personal information is most often used in CCPA (the California Consumer Privacy Act). CCPA defines PI as "Information that identifies, relates to, describes, is capable of being associated with, or could reasonably be linked, directly or indirectly, with a particular consumer or household." Personal information is a superset of PII. Simply put, all PII is considered personal information, but not all personal information is considered PII.

In this book, I'll use 'personal data,' but both 'personal information' and 'personal data' are used interchangeably and can be used without distinction, but the term personal 'data' is mainly used in terms of utilization and personal 'information' in terms of protection in Korea. Korea's 「Personal Information Protection Act」 defines personal information as 'information on living individuals', 'information that can identify individuals through names, resident registration numbers, videos, etc.' or 'information that can be easily combined with other information even if a specific individual cannot be recognized by itself'. The GDPR states, 'Personal data are any information related to an identified or identifiable natural person.' An identifiable natural person is defined as a person who can be identified directly or indirectly by referring to an identifier such as a name, identification number, location data, online identifier, or one or more factors related to physical, physiological, genetic, mental, economic, cultural or social identity characteristics.

The first of the four components of personal data definition defined in the June 2007 'Opinion 4/2007' by the Working Party 29 on the concept of personal data is 'any information'—all information related to an individual, regardless of its nature, content, and format, is personal data. The second is 'relating to'. The third is whether it is 'an identified or identifiable.' The fourth is the 'natural person'. This means that the information of the deceased is not personal data. However, in the case of death data, it is partially different from country to country but tends to be covered by medical law. Except for the fourth element, it is intentionally defined to enable an extensive range of interpretations. In particular, the third factor, identification, is likely to eventually lead to a sharp confrontation between the industry and individuals. The reason is that the criterion is 'identifiability.' Conservatively speaking, even a minimal probability should be defined as personal data if identification is possible. Even if identifiers are deleted and de-identified, several re-identification methods are applicable, and the likelihood of re-identification varies depending on the composition of the population data. Of course, the extreme level of insignificant possibilities will be excluded at the practical or legal level. In fact, when defining personal information in the revised 'Personal Information Protection Act' in Korea, it stated, "Information that can be easily combined with other information even if it cannot recognize a specific individual. In this case, the time, cost, and technology to identify an individual, such as the availability of other information, should be reasonably considered. In other words, if the data is combined and re-identified with unreasonably large amounts of time, cost, and technology, the data should be regarded as anonymous rather than personal data. On the one hand, it seems to be a reasonable standard, but the trap is that there is no standard for rationality. Nevertheless, the broad definition of personal data remains steady in the GDPR, so this view in WP29 is essential for understanding the meaning of personal data. The European Union authorities aimed to broadly define personal data to include all information about identifiable individuals. This definition is far broader than data infringement laws in many states in the United States. Based on this, personal data includes a considerable range of information. The same is true even when there is a slim connection between such information and identifiable individuals. The reason for maintaining such a conservative definition is that privacy is, after all, a matter of human rights.

Ownership of Personal Data

The definition of big data itself has nothing to do with whether or not personal data is included, but in practice, big data is mainly in the form of non-personal data. The reason is the legal logic of consent for personal data. The personal data collected by companies is necessary for the provision of services, and consent must be obtained to use personal data for a purpose other than the purpose of providing the service. However, the use of big data is ultimately aimed at getting new insights through data analysis, and consent from individuals cannot be obtained in advance because analysis must be started without knowing what the insights are. Since the insight to be verified cannot be about a specific individual, it is challenging to conduct a direct analysis of the individual through big data. In addition, under the premise of compliance, the big data set has practical limitations that make it challenging to include personal data such as identification information.

There are two significant privacy-related risks in using big data that are basic in all industries. First, even if it is big data consisting of de-identification processing personal data, the possibility of re-identification increases as the amount of data itself increases. The second is that each piece of data may not be very sensitive, but it is possible to infer personal sensitive information unintentionally. Technically, it is easy to conflict with the value of privacy at this point because one of the goals of big data analysis is to gain insights about users and people and create new services and values based on them for innovation. Insights about users can be obtained through big data or machine learning, but there is a demand to identify and target individuals profiled through insights to reach those users. And this demand eventually leads to a data combination between heterogeneous industries.

Personal data ownership is not the same concept, although it will be expressed as 'ownership,' the same term in civil law. Since data should be copied, reproduced, and shared indefinitely, it should be introduced as a concept of data ownership that fits the nature of the data, not as an exclusive or exclusive ownership concept. The EU is also discussing the premise of the concept of ownership unique to non-exclusive data. The white paper "Building a European Data Economy" suggested that "data ownership" should be introduced as a new right to build Europe's data economy. Some believe that copyright laws can protect the right to own, but in the case of data without creativity, it is difficult to protect it by copyright. Privacy Acts in many countries also protect personality rights, and it is difficult to regard them as protection in terms of property rights. In reality, data is being traded on the market, and no one denies that it has property value. Legally, the concept or scope of the use of data or the rights and obligations between the parties involved can be sufficiently regulated under the contract law, but legal uncertainty may persist. Contracts are relative and flexible for each party, so they are very insufficient as legal protection measures, and the value of the data depends heavily on the ability to negotiate, which only adds to the uncertainty. In that case, the data market and economy cannot grow because there is no stable legal protection for data value.

Fortunately, it is expected that a slightly different approach will be possible to personal information other than general information. At least because there are legal means of protection for personal data. In advance, this book tries to address the proposal of ownership of access to personal data. The access authority has several advantages because it can be an exclusive right. In particular, regarding ownership, discussions become easier in data combination scenarios. This is because it is difficult to oppose the idea that an individual has all the authority over their data.

III.

Data Sovereignty
and GDPR of the European Union

Background

An invisible war between advanced IT countries to protect data sovereignty–the country's sovereignty–is fierce. It is no coincidence that global IT companies are the main characters in the U.S.-China trade war. As the trend of all personal information, including netizens' political orientation and economic activity trends, is strengthened, countries are protesting against this. We stand at the crossroads of whether global economic hegemony remains in the United States, and the determining factor is who secures more data. China appears to have a desperate duty to take global economic hegemony from the United States or at least weaken its power. China's Network Safety Act prohibits the transfer of domestic data abroad starting in 2019. All IT companies operating in China must keep their data in China. If requested by the Chinese government, data decryption information should also be provided at any time. In other words, it is not a barrier that prevents people from falling outside but a barrier that prevents people from going out from inside. Eventually, Apple also managed iCloud's data in China through a Chinese state-owned company. This is actually a cyber trade barrier. Data protection is essential when utilizing, and data use in good faith can only be achieved when data protection is guaranteed. In the entire process of production, processing, and consumption of personal data for the protection and utilization of personal data, sovereignty over personal data is, after all, an individual right and duty.

Data Protection and Utilization

Usually, data protection and utilization may not be easy to achieve together, but ultimately, they are goals that cannot be abandoned. Who is responsible for protecting personal data needs to be reviewed together in terms of overall governance, including data utilization, and it is natural that this also leads to data ownership discussions. During a US Senate hearing, Facebook CEO Mark Zuckerberg said several times that on Facebook, 'the user owns the data.' According to a survey conducted by Insights Network, 79% of consumers said they want a reward for sharing their data. In other words, companies operating services and individuals receiving services perceive data as subject to 'ownership.' Some policymakers, such as Senator John Kennedy, have advocated through legislation such as the Own Your Own Data Act of 2019 that 'Each individual owns and has an exclusive property right in the data that individual generates on the internet' and required social media companies such as Facebook to obtain licenses to use this data. California Governor Gavin Newsome) has proposed a data dividend designed to share wealth generated from personal data. In Korea, data dividend was experimentally implemented in 2020 in Gyeonggi Province. Data dividend is the concept of returning profits to individuals who contributed to data production in relation to profits generated by companies' collection and use of personal data. Gyeonggi Province earned 100 million KRW from 11 data sales revenues from 2019 to 2022. Though it was a small amount of money, the dividend itself is based on ownership. Fundamentalist personal data thinkers like Alan F. Westin have defined privacy as an individual's right "to control, edit, manage, and delete information about them[selves] and decide when, how, and to what extent information is communicated to others."

However, whether guaranteeing ownership is desirable in protecting personal data and privacy remains controversial. In terms of actual protection, the data is not the problem. Treating data as if it were property alone does not make individuals continue to pay attention to the value provided by and to their personal data, even if individuals choose to 'sell.' Data is not just a commodity. There is a demand for flows because our personal data is valuable to businesses and organizations for many purposes, and a set of systems for all rights, such as patents, copyrights, and other intellectual property or privacy rights, remain tense about the free flow of data.

Interests Surrounding Personal Data

For example, a name is the most essential part of a person's identity and is personally identifiable information protected by privacy acts in many countries. While most of us are unaware, our name is how others perceive us and is used in society for friends and family and who we are and for voting lists, property registers, financial accounts, and numerous other social and economic contexts. Names are also essential social and economic resources that are valuable to others. It is because of this social and economic importance that naming is regulated. A name is assigned at birth and registered on the birth certificate, and any name change requires formal procedures and approval. In this way, personally identifiable information is a personal and social tool.

Other personal data that we must share with our names includes cross-interests. The transactions we conduct through our banks are also part of the bank's business records. However, the bank's interest in these records does not necessarily mean it has absolute control to do what it wants with this data. For the most part, the data we generate is shared more broadly in addition to the first collector, but even in this case, stakeholders' interest in the data is not exclusive. The digital economy operates in a wide range of data-sharing ecosystems used for various purposes, including cases such as credit survey reports. This perspective is also reflected in the EU's GDPR and CCPA in the U.S..

As companies increasingly start outsourcing services and moving data processing to cloud services, more personal data is stored in places that are unknown to individuals. The data stored here is sometimes used by third stakeholders. For example, in the case of advertising and marketing, we recognize tracking and data sharing due to visibility to consumers but share information without our recognition because mechanisms of advertising technology, such as advertising identifiers and cookies, can track sites and users' devices. And data sharing also contributes to vital public interests and values. Therefore, what is needed for individuals and society is to ensure that personal data that is useful to social, economic, and government systems is available while protecting each of our important interests in our personal data.

Consumer Data Sovereignty

The industry discusses consumer data sovereignty to increase consumer profits in the data economy era. Consumer data sovereignty is usually the exclusive right of the consumer, the data subject, for the creation, storage, distribution, and utilization of consumer data and is the right to control the flow, disclosure, or non-disclosure, use, etc., of data for the consumer's benefit. The problem is that there is a difference between the right to self-determination of personal data and the basic direction of approach. The right to self-determination of personal data is treated as a basic right based on the constitution and a general personal right. It is understood as a part of human rights in many countries. On the other hand, there is a fundamental difference in that the subject of consumer data sovereignty is limited to 'consumers.' Since passive situations like receiving personalized, customized products and service recommendations using advanced analysis technologies such as machine learning and artificial intelligence are the main scenarios, it seems to represent consumer interests with consumers' exclusive rights. But there is a big difference in how they view privacy.

There is also an argument for moving away from the form of obtaining consent from individuals to use it more actively for now and to provide transparency about data processing afterward. If the consent process is complicated, it is argued that personal data cannot be protected because it is almost unconsciously agreed upon. It is more effective to guarantee the right to follow-up data processing. However, this is a very business-friendly position, and once an accident occurs, it may be a useless argument for an individual who has suffered. Privacy can suffer damage that is difficult to recover from even a single invasion. Personal data is like fire, so if we use it well, it enriches us, but if it is misused, it can take everything away. We can't put people who say they want to be comfortable at risk because it's a matter of human rights, the universal value that supports this society, after all.

A good example of a conflict of values is the backdoor controversy between the FBI and Apple around February 2016 to unlock devices. In the wake of a gun attack in San Bernardino, California, that killed 14 people, the FBI asked Apple to remove the security function of the iPhone used by the terrorist. However, the controversy erupted when Apple refused to do so, claiming it violated privacy. The FBI argued that it was a legitimate public power of state agencies and could issue a cooperative order based on the "All Writs Act" introduced in 1789. Apple protested against this, saying it violates freedom of expression and the right not to be infringed unfairly. Eventually, it spread to a legal battle, and even among U.S. federal courts, opinions were divided between "the iPhone should be unlocked" (a California court) and "there is no obligation to release" (a New York court in New York). The FBI has dropped the lawsuit in search of ways to decrypt without Apple's cooperation. However, it is a good example of what logic works when privacy values and other values collide, and it is hard to explain simply by the logic of consumer data sovereignty.

Introduction of GDPR (Scopes and Principles)

The GDPR has become a de facto global standard for personal data protection, despite criticism that the extension of the application of the law to EU residents was aimed at blocking cross-border transmission to protect the EU's digital single market under the pretext of privacy protection in terms of human rights. Personal data under the GDPR means 'any information related to an identified or identifiable natural person.' An identifiable natural person is defined as a person who can be identified directly or indirectly by referring to an identifier such as a name, identification number, location data, online identifier, or one or more elements related to physical, physiological, genetic, mental, economic, cultural or social identity characteristics Among personal data, certain types of data are sensitive data that can pose significant risks to an individual's basic rights and freedoms and require special management. These include ethnic or ethnic origin, political views, religious or philosophical beliefs, union membership and genetic information processing, biometric data to uniquely identify natural people, and natural people's sexual life or sexual preferences.

GDPR inherits the concept of controller and processor defined in the previous Directive (Directive 95/46/EC). Under the GDPR, a controller is defined as 'the natural or legal person, public authority, agency or other body which, alone or jointly with others, determines the purposes and means of processing personal data.' In contrast, processor means 'a natural or legal person, public authority, agency or other body which processes personal data on behalf of the controller'. The GDPR applies to organizations 'established' within the EU and organizations outside the EU that sell products or services to or monitor individuals within the EU. The establishment is that any processing of personal data relating to the activities at the premises of a controller or processor within the EU is subject to GDPR, regardless of whether the processing takes place within the EU. There is no definition of establishment in GDPR. Still, GDPR is applied very widely in practice since activities are also unrelated to whether the processing is carried out within the EU. For an extreme example, if Samsung Electronics (regardless of its headquarters location) collects personal data generated while a Korean, a Samsung Electronics customer, resides in Paris, it will be subject to GDPR.

Data processing principles in GDPR include lawfulness, fairness and transparency, purpose limitation, data minimization, accuracy, integrity, and confidentiality.

The principle of lawfulness is that the data controller should only process personal data where there is a legal basis for processing the data, which is the consent of the user, the performance of the contract, legal obligations, vital interests of the individual, the public interest and legitimate interests pursued by the controller or by a third party. Fairness means that, in addition to being lawful, the processing of personal data must be fair, and transparency means that when processing personal data, the controller must be open and clear about the data subject.

The purpose limitation principle states that personal data should be collected and processed for specified, explicit, and legitimate purposes and not further processed in a manner that is incompatible with those purposes. The principle of data minimization means that personal data is adequate, relevant, and limited to what is necessary in relation to the purposes for which they are processed. Accuracy means that personal data is accurate and, where necessary, kept up to date; every reasonable step must be taken to ensure that inaccurate personal data, regarding the purposes for which they are processed, are erased or rectified without delay.

The storage limitation principle is that personal data

must not be kept in a form that permits identification of data subjects for no longer than is necessary for the purposes for which the personal data are processed. The integrity and confidentiality principle is to process personal data in a manner that ensures appropriate security of the personal data, including protection against unauthorized or unlawful processing and accidental loss, destruction, or damage, using appropriate technical or organizational measures.

Based on these principles, the GDPR also explicitly presents the standards for lawful processing. First, the data subject has given consent to processing their personal data for one or more specific purposes. At this time, consent must be freely given and specific, sufficient information must be provided, and the consent and wish of the information subject must be explicit. Thus, making consent impossible, giving an abstract explanation, or selecting a checkbox in advance, even if an individual consents, is likely invalid.

Lawful processing also includes cases necessary for the performance of a contract to which the data subject is a party or to take steps at the data subject's request before entering into a contract. In addition, when it is necessary to fulfill the controller's legal obligations, to protect the important interests of data subjects or other natural persons, and for the public interest and the legitimate interests pursued by the controller or third party, those are supposed to be lawful processing.

The GDPR provides more active protection by prohibiting

the processing of personal data revealing racial or ethnic origin, political opinions, religious or philosophical beliefs, or trade union membership, and the processing of genetic data and biometric data to uniquely identify a natural person, data concerning health or data concerning a natural person's sex life or sexual orientation.

Rights and Obligations for Data Protection in GDPR

Fundamentally, providing information to data subjects is a key component of GDPR. It is important throughout the data protection framework, from its impact on fairness to its ability to rely on legitimate interests for processing. However, there are considerable limitations in complying with GDPR's obligations related to transparency in the mobile environment, especially in most companies that provide mobile-based fintech services.

Under GDPR, the enforceable rights of data subjects against companies, institutions, and organizations that process their data are extensive. Articles 12, 13, and 14 stipulate the right to receive transparent communication and provision of information, and Articles 15, 16, and 17 stipulate the right to access, rectify, and erase personal data. Articles 18, 19, and 20 describe the right to request restriction of processing of personal data, the right to be notified, and the right to data portability. Articles 21 and 22 clarify the right to object to the processing of personal data, including the right not to be subject to automated decision-making, including profiling.

Among these, in particular, in the case of the right to data portability, there is an aspect of strengthening the data subject's rights. Still, there is also an aspect of inducing fair competition between companies. It is provided in a machine-readable form while dealing with the concept of PDS, which has influenced the advent of the recent MyData industry. The right not to be subject to automatic decision-making significantly impacted the AI regulation proposed by the EU's Commission in April 2021. The AI regulation plan aims to be a flexible and balanced framework so that excessive restrictions do not hinder technology development while specifying high-risk situations brought by artificial intelligence. This was proposed as a draft for the first legally binding bill related to artificial intelligence. On 14 June 2023, MEPs (Members of the European Parliament) adopted the Parliament's negotiating position on the AI Act.

Transparency in how personal data is used, i.e., the requirement for openness and honesty, is a major component of GDPR. Ultimately, GDPR aims to make it clear to data subjects that they know their rights, risks, rules, and protective measures about the processing of personal data. When collecting personal data from the data subject, it is obligated to provide information to the data subject. Because fair processing information, including electronic means, can be provided in writing, notifying individuals that they have processed fairly will be one of the convenient ways for controllers to comply with the transparency requirements of GDPR. It also clarifies the controller's obligations by specifying various requirements to be fulfilled, such as policy implementation, training, documentation, designation of DPO (Data Protection Officer), DPIA (Data Protection Impact Assessment and BCR (Binding Corporate Rules).

Data Processing and Protection Measures in the GDPR

One of the explicit objectives of the GDPR is to allow the free flow of personal data within the EU by the consent-based data protection principle. At the same time, however, the GDPR recognizes that transferring personal data to third countries requires special consideration. Although GDPR does not define the concept of data transfer, data transfer is different from simple data transmission. It is a concept that includes processing within a third country that has completed a 'transfer.' Therefore, even if personal data is transmitted through a third country on the way from an EEA (European Economic Area) country, the transfer of such data is not within the scope of the GDPR unless substantial processing of the personal data is carried out in a third country. As for the data transfer, procedures should be in place to provide an appropriate level of protection and to designate a country with appropriate protection. Safe Harbor, Snowden Effect, Safe Harbor II, and Privacy Shield in the U.S. are all about providing appropriate safety measures. Such measures include codes of conduct and certification mechanisms. The most significant development in international data transfer under GDPR is the inclusion of BCR (Binding Corporate Rules) as a mechanism that can be used for both controllers and processors to legalize such transfers within the company group. Of course, one of the advantages of GDPR is the flexibility that the transfer of personal data can still be performed if it is one of the exemptions for certain situations included in the GDPR, in the absence of an adequate level of protection or adequate protection, which is one of the reasons why the basic basis for balancing the use of personal data while focusing on the protection of personal data is accepted as a practical international standard of GDPR in terms of regulatory law.

Article 22 and Recital 71 in GDPR state the data subject should have the right not to be subject to a decision, which may include a measure, evaluating personal aspects relating to them which is based solely on automated processing and which produces legal effects concerning them or similarly significantly affects them, such as automatic refusal of an online credit application or e-recruiting practices without any human intervention. Such processing includes 'profiling' that consists of any form of automated processing of personal data evaluating the personal aspects relating to a natural person, in particular, to analyze or predict aspects concerning the data subject's performance at work, economic situation, health, personal preferences or interests, reliability or behavior, location or movements, where it produces legal effects concerning them or similarly significantly affects them.

As the number of services incorporating artificial intelligence increases, the importance of personal data related to this is also increasing. Therefore, in the early stages of introducing artificial intelligence services, it is necessary to thoroughly analyze personal data issues caused by artificial intelligence and transparently disclose the personal data process flow so that users' trust is not lost in the future.

Data Combination

As mentioned in the previous chapter, the demand eventually leads to a data combination between heterogeneous industries. But, in the case of the EU or the United States, there are no more explicit regulations on data combinations than expected. In GDPR Article 4, 'processing' means any operation or set of operations that are performed on personal data or sets of personal data, whether or not by automated means, such as collection, recording, organization, structuring, storage, adaptation, or alteration, retrieval, consultation, use, disclosure by transmission, dissemination or otherwise making available, alignment or combination, restriction, erasure or destruction. Article 35 states that where a type of processing, in particular using new technologies and taking into account the nature, scope, context, and purposes of the processing, is likely to result in a high risk to the rights and freedoms of natural persons, the controller shall, before the processing, carry out an assessment of the impact of the envisaged processing operations on the protection of personal data, which we call it as DIPA (Data Protection Impact Assessment). The personal data controller can secure the basis for proving that appropriate measures have been taken to comply with the GDPR. The GDPR exemplifies three examples of personal data processors that can cause high risk. The personal data controller and processor can secure the basis for proving that appropriate measures have been taken to comply with the GDPR. The GDPR exemplifies the following three examples of personal data controllers and processors that can cause high risk through Article 35.

> 3. A data protection impact assessment referred to in paragraph 1 shall in particular be required in the case of:
> (a) a systematic and extensive evaluation of personal aspects relating to natural persons which is based on automated processing, including profiling, and on which decisions are based that produce legal effects concerning the natural person or similarly significantly affect the natural person;

(b) processing on a large scale of special categories of data referred to in Article 9(1), or of personal data relating to criminal convictions and offences referred to in Article 10; or

(c) a systematic monitoring of a publicly accessible area on a large scale.

However, the DPIA guidelines issued by Working Party 29 point out the following nine factors to consider when determining high risk.

1. Evaluation or scoring
2. Automated-decision making with legal or similar significant effect
3. Systematic monitoring
4. Sensitive data or data of a highly personal nature
5. Data processed on a large scale
6. Matching or combining datasets
7. Data concerning vulnerable data subjects
8. Innovative use or applying new technological or organisational solutions
9. When the processing in itself "prevents data subjects from exercising a right or using a service or a contract"

The guidelines of Working Party 29 generally recommend that DPIA be performed if two or more of these nine fall under. Considering the sixth factor, data combination is eventually recognized as one factor determining whether it is a high risk among various types of data processing.

There can be several combinations of data combinations between heterogeneous industries. The financial sector can be divided mainly into banks, cards, securities, and insurance, and the types of heterogeneous industries that the financial sector is cooperating with can be classified into tele-communications, distribution, game, and medical industries. In addition, although not classified as an industry, platform companies are also subject to data cooperation. Industries have different characteristics; accordingly, the types of data collected from each industry differ. To establish various heterogeneous industries and data cooperation strategies in the financial sector, it is necessary to organize the characteristics and goals of cooperation between the financial sector and each heterogeneous industry. Examples of combinations centered on combination with financial data are as follows.

The first is the combination of financial data and communication data. When the financial sector and the telecommunications industry cooperate, it is characterized by its strength in combining media content. In addition, it is easy to analyze the characteristics of the commercial district by combining card usage history and location data. Through this, it can aim to combine media content and financial services, sell data-combined products, and develop new business models.

Secondly, it is the combination of financial data and distribution data. In the case of the distribution industry, it has customer purchase data from online and offline stores. It can be used for customized marketing by combining purchase data with consumption pattern data collected through the financial sector. Through data cooperation with the distribution industry, the financial sector aims to make it easier to collect, share, combine, and utilize data collected through various channels through data platforms. Furthermore, cooperation to carry out joint MyData projects based on these platforms continues.

Third is the combination of financial data and game data. It can be seen that the financial sector is trying to implement strategies to maximize the use of games and artificial intelligence technologies and target the MZ generation through cooperation with the game industry. In some cases, the company aims to establish a new type of digital securities company by utilizing the technology possessed by game companies. It can develop content that links finance and games and use them for marketing.

Fourth, it is the combination of financial data and medical data. Through cooperation with the medical industry, such as hospitals, lifestyle data using medical information and wearable devices can be used. Based on this, a MyData platform can be configured, and customized insurance and medical services can be provided to maximize customer convenience.

The fifth is the case of collaboration between financial institutions and platform companies. The financial sector can easily use existing platforms when cooperating with platform companies. Contact points can be created with customers of each platform, and the characteristics of each platform can be utilized. The financial sector can build banking services inside the platform and generate profits through customized financial services and additional services using the platform.

The demand for data from companies and institutions is increasing, and new businesses using data that combine various data are also becoming more active. In addition to these changes in the financial industry environment, consumers are becoming more aware of data sovereignty, and their need for personalized, comprehensive financial and living services is increasing. The value created when various data are combined must be utilized to respond to the changing industrial environment and meet these customer needs. To this end, close cooperation between heterogeneous industries should be continuously sought, and data cooperation strategies should be prepared. Of course, security and privacy should be fundamental in this process.

Data Sovereignty on Combined Data

Ownership of combined data has not been discussed in depth in academia or industry. Before discussing the sovereignty of combined data, it is necessary to organize case by case whether it is privately owned, owned by the source data holding institution or combining institution, or jointly owned. If personal data is de-identified or anonymous, it is converted to non-personal data, so in principle, it is not subject to the privacy acts of the countries. However, not only is there ambiguity in the criterion for determining identifiability itself, but even if identifiability is removed, there may be re-identifiability over time due to the increase in combined information, the expansion of access to combined data, and the advancement of analysis techniques. Therefore, even if personal data is de-identified or anonymized, additional measures are still needed to block the risk of re-identification. This is the same when personal data is provided by pseudonymization and even when de-identified data is combined. There is a risk of being evaluated as a combination by a pseudonym identifier. Suppose each data provider "de-identifies" the data set, replacing only the identifier with a matching key and providing data without removing the uniqueness of each data. In that case, this data is pseudonym data, and each data provider gives it to a trusted third party with both the original data and the data converted to the matching key. Hence, the above provision is a third-party provision of personal data. In other words, de-identification at this stage cannot be called de-identification in the legal sense. A combination of data in the form of a unique matching key and a reliable third party remains at risk of being evaluated as illegal unless there is a separate legal basis.

The combination of non-personal data does not cause special legal problems, whether the data provider or the data recipient. When a data provider does, data is created within that range and only has the nature of a contract. On the other hand, the combination of data, which is personal data, not only processes its own personal data but also increases the amount of information, further increasing the identifiability. There is no legal problem with the data subject's consent, but there are several legal problems in performing it without the data subject's consent. However, on the other hand, the value of data is often maximized in identifying the connection relationship by combining data collected at different levels.

The distinction between personal and non-personal data is a relative concept that depends on various circumstances, such as the possession and accessibility of combined data. Therefore, in the case of generating and providing combined data, the legality of processing is secured only when each party blocks the possibility of accessing their or the other party's combined data. In particular, if there is a risk of re-identification if there is a combination of data due to the lack of anonymity, it is necessary to block any contact and information exchange that increases the risk of re-identification, in addition to the exchange of non-identification data, including matching keys between each data provider and a trusted third party to perform the combination and obligate the data recipient to take measures such as suspending additional use and discarding or reprocessing if the risk of re-identification occurs or increases. It is the same as if there is one data provider. Still, three or more people are involved in data combination, so each responsibility in the data contract should be clarified.

In the case of data combination, personal data protection issues become a complicated problem to deal with. In this situation, social consensus on data combination and its utilization benefits become more complex, eventually leading to discussions on data sovereignty. This area requires many academic and legal discussions in the future, but commercially, other approaches are needed. Otherwise, while there is no social consensus, the industry will remain uncertain, which will not help humanity's prosperity.

IV.

MyData and the Changes It Will Bring

Background

MyData can also be defined as a series of processes or a movement that aims to give individuals the full right to make decisions about their personal data. It is a concept that seeks to strengthen individuals' right to decide on their personal data. It is important to have a broad understanding of internationally accepted concepts, principles, standards, and examples of MyData. Companies can use data to develop new products and services, identify customer needs and provide customized services, increase corporate productivity through data analysis, and innovate business models. It is also expected to bring about major changes in society and politics in the data economy era. For example, the government will be able to open public data to improve the quality of life for the people and increase national competitiveness. In addition, it can be used to understand the direction of the cultural industry through big data analysis and to develop new cultural content.

MyData is often introduced first into the financial sector. To complete the 4th industrial revolution in the financial industry, providing low-cost customized services through online/mobile platforms is essential. To perform these services properly, MyData must be adopted to grasp customers' financial status and needs accurately. A different approach is required than customer modeling on YouTube, Netflix, or general e-commerce. Even if the recommendation of video, music, and products does not match the customer a little, there is no significant loss to the customer. Still, in financial services, the wrong recommendation can cause a big loss to the customer. Therefore, numerous studies are essential to analyze financial conditions and needs based on individual financial activities. Numerous studies must be conducted from various angles to be the foundation and provide an appropriate service to actual customers. Based on the results of multiple clinical studies, it is in a similar context to medical services that perform prescriptions and treatments suitable for patients as much as possible. Suppose it is based on various financial activity data of hundreds of millions of people. In that case, it will be possible to accurately identify the types and needs of customers in the classification of their financial status and much more detail. Numerous true customized financial services can be created if they are well utilized. Of course, this requires thorough preparation for privacy protection because the customer's financial activity data contains very sensitive personal information, and protocols should be in place to consistently integrate data collected and managed in different ways by various financial companies.

MyData? Personal Data?

As written, 'MyData' is my data and basically personal data. However, not all personal data is called MyData. The reason is that MyData encompasses both the conditions and a series of processes necessary for an environment in which an individual has complete control of self-information, not the data itself. Therefore, the MyData Global Association explains MyData by declaring the goals and principles pursued by the MyData concept instead of defining it separately. The purpose of MyData is to return the decision-making power over personal data to the parties so that they and their societies can develop knowledge, make more informed decisions, and communicate more initiatively and efficiently between individuals and organizations. It takes the form of a declaration to join the movement to change the paradigm of personal data by publicly declaring and insisting on the value pursued by MyData, and what it is trying to change is listed as follows.

The first is to make 'formal rights into actionable rights'. Many individuals are unaware of their rights and find it difficult to exercise them, and they have usually been maintained only formally, covered by corporate practices. Quite a few individuals still subconsciously check "Check" and agree when they do not properly understand the requested consent. MyData aims to ensure that consent based on true transparency and accurate understanding becomes a standard for communication between people and organizations. It aims to make it a so-called 'one-click right' that is simple and efficient to implement access to personal data, rewards, mobility, and the right to be forgotten.

The second is data protection and self-determination. Although regulations on protecting personal data and corporate ethics are in place in most countries, it is still difficult to say that there is a sufficient guarantee of self-determination in using personal data held by companies and organizations. Even the EU Member States that have introduced the GDPR. Non-EU countries had to sign Standard Contractual Clauses to transfer personal data collected from the EU. GDPR compatibility reviews and local administrative procedures require up to a year and hundreds of millions of dollars. Ultimately, it took unnecessary money and time to overcome this situation because it did not guarantee personal data protection and self-determination in GDPR, which is now a de facto international standard.

The third is the construction of an open ecosystem. Today's data economy has become a monopoly market dominated by a handful of large companies and platforms. These companies and platforms cannibalize the market, blocking direct access to customers by many SMEs (Small and Medium-sized Enterprises) and competitors. MyData allows individuals to control the circumstances that occur with their data, allowing individuals to decide and provide unobstructed data anywhere in the world freely. It aims to create a truly free flow of data.

MyData Principle

The principles of MyData are presented in the declaration of the MyData Global Association. It is essential that individuals not only know and control their personal data but also gain insights from it and be able to claim their share of the profits generated from using it. In particular, in collecting and trading personal data and making decisions based on personal data, companies, and organizations have an absolute power to decide than individuals. Accordingly, the MyData Global Association presents the following principles regarding balancing power.

The first is human-centric control of personal data. Individuals should be empowered actors in managing their personal lives both online and offline. They should be provided with the practical means to understand and effectively control who has access to data about them and how it is used and shared. Privacy, data security, and data minimization should be standard guidelines when designing applications. Companies and organizations should inform individuals of their privacy policies and how to activate them. The principle is that individuals should be authorized to give, deny, or revoke consent to data sharing on the premise that they clearly understand why, how, and how long their data is used, and ultimately, terms and conditions for personal data use should be traded fairly between individuals and organizations.

The second is that the individual should be the "hubs" of personal data (Individual as the Point of Integration As personal data becomes more diverse, the threat of invasion of privacy also grows proportionally, which can be resolved if an individual becomes a kind of "hubs" where cross-references of personal data occur.

The third is the principle that individuals should be considered free and autonomous agents capable of setting and pursuing their own goals. They should have agency and initiative. The principle is that individuals safely manage their personal data how they want and receive tools, technologies, and support to make autonomous decisions based on their personal data.

The fourth is the principle of 'Data Portability,' one of the most frequently mentioned issues related to MyData. It ensures that individuals obtain and reuse their data for their own purposes or other services. The right to portability of personal data is the key to transforming data in closed storage into valuable data as a resource that can be reused. This means that it must be combined with practical means, such as allowing individuals to effectively move their personal data, both for download to their devices and for transmission to other services. It is a principle that data holders should provide such data in a structured, generally available, safe, and easy-to-use machine-readable format.

The fifth is the principle of 'Transparency and Accountability.' Companies and organizations that use personal data must transparently inform individuals what they do and why with the data and are responsible for keeping what is announced. Individuals should be able to hold companies and organizations accountable for unintended consequences, such as security accidents and intended consequences caused by storing and using personal data.

The sixth is 'Interoperability.' The principle is to eliminate the possibility of data lock-in to specific companies and organizations while reducing friction caused by exercising the right to move data. Interoperability can be achieved by standardizing business procedures and technologies, which requires ongoing efforts for data interoperability, open APIs, protocols, applications, and infrastructure. It also enables movement and reuse while maintaining individual control over personal data.

The purpose and principles of MyData are well introduced on the MyData Global Association site. Ultimately, the MyData spirit and movement may eventually depend on how civil society is aware of data sovereignty, which can differ by country.

Changes that MyData will bring

The most significant change that individuals will experience through MyData is that individuals will become active subjects exercising full rights to their data. Individuals are no longer passive beings who need to receive personal data protection services. Another is the change in the data usage environment. It changes from a centralized and closed type centered on data dinosaurs to a distributed and open ecosystem. 2017 MyData Declaration states access and redress, portability, and the right to be forgotten to become "one-click rights": rights that are as simple and efficient to use as today's and tomorrow's best online services. This is the declaration of the transition from a formal right to an actionable one.

The second change described in the MyData Declaration is the shift from data protection to data empowerment. This is a new perspective that sees people not just as passive beings to be protected from harm but as full citizens and their agents, willing and able to improve their lives through the use of collected personal data.

The third change that MyData wants to realize is the transition from a closed ecosystem to an open ecosystem. The rapid growth of the data economy is based on network effects. Currently, the most influential users of personal data are platform companies with the ability to collect and process large amounts of personal data. These platforms are aimed at lock-in and have closedness in themselves. MyData is focused on placing people at the center of data governance, enabling free data flow, and creating balance, fairness, diversity, and competition in the data economy ecosystem.

With the advent of the data economy era, companies can use data to develop new products and services, identify customer needs and provide customized services, increase corporate productivity through data analysis, and innovate business models. It is also expected to bring about significant changes in society and politics in the data economy era. For example, the government will be able to open public data to improve the quality of life for the people and increase national competitiveness. In addition, it can be used to understand the direction of the cultural industry through big data analysis and to develop new cultural content. The basic purpose of the "MyData" project is to guarantee individuals who are data providers the right to self-determination of their data and to allow them to legally and technically control and manage their personal data. Through this, it aims to promote the vitalization of the data economy while enhancing the people's benefits and quality of life.

In summary, individuals can make decisions about accessing and using their data and receive fees or benefits in exchange for providing it. In addition, financial institutions will be able to use various data to identify customer needs and tendencies and provide differentiated financial services at any time if customers want. Ultimately, the beginning of MyData is a vital signal to break down the boundaries between industries and signal the start of the era of a full-fledged data economy. In addition, as the amount of data explodes, there is a growing recognition that combining financial and non-financial data provides a more comprehensive understanding of individuals' and businesses' financial positions and behavior. Financial institutions can combine non-financial data sources such as customers' social media activities, consumption behavior, geographic location data, and even biometric information to develop services from a more diverse and detailed perspective on customers, providing customized financial products to meet their needs and preferences, such as credit rating and identification of potential investment opportunities.

A representative example of this trend is the emergence

of an alternative credit rating model. It is a method of evaluating the creditworthiness of individuals with limited or no credit records by converging non-financial data. This is because factors such as online shopping habits, social relationships, and even smartphone usage history can provide valuable insights into borrowers' willingness and ability to repay their loans. Moreover, the convergence of financial and non-financial data is not limited to lending decisions, and for investment companies, alternative data such as satellite images of investment companies, customer reputation and social sentiment analysis, and supply chain data are increasingly being used to gain a competitive advantage in the financial market. Using data from a non-financial perspective can provide new indicators of market trends and corporate performance, helping investors make investment decisions based on more information. In recent years, as such analyses gain traction, new financial service models that utilize the power of combined data are emerging one after another. Examples include customized consulting based on individual lifestyles and preferences. This pre-fraud detection system analyzes financial transactions, user behavior, and insurance products that respond in real-time to market changes.

However, the reality is that financial institutions still lack experience in data-based customer service. Therefore, various experiences are gained through the convergence of financial and non-financial data to expand innovative financial services. Along with this experience, there is another important task for the financial sector to prepare for. Financial companies should prepare internal management guidelines and systems according to the financial supervisory authority's regulatory guidance and legislation introduction. Preparing AI ethics guide policies to use data responsibly and ethically is also necessary. Because the convergence of financial and non-financial data can raise privacy issues, balancing this innovation with consumer protection is essential for regulators and industry participants.

Actionable Rights and Open Ecosystem with Data Empowerment

The right to control data about oneself in the digital age must be regarded as a human right. This is because we do not exist as a body in the digital world but as data. In terms of world history, personal data protection laws have their roots in the Universal Declaration of Human Rights adopted by the United Nations General Assembly at the time after World War II. The MyData Principles are also one of the ways to guarantee people these rights and provide practical tools to exercise them. Control over personal data begins with recognizing the nature and extent to which personal data is collected. Individuals should have access to that data and be able to modify or delete the data if necessary. They should be able to identify who is using the data, share it for whatever purpose they want, and stop such sharing. This does not mean that this authority is not absolutely restricted. An individual's right to control personal data can sometimes be limited by law or other means, just as having the authority to control a car does not mean they have the right not to comply with traffic laws.

However, simply following the requirements of the law is not enough to realize the MyData model and build a human-centered personal data ecosystem. Changes in laws, regulations, and technology can all contribute to the realization of MyData. Regulations can accelerate or slow change, but legislation alone cannot bring about change. Well-organized law does not mean that individual rights are better guaranteed or that industrial use of data is not guaranteed. MyData aims to make personal data a usable resource premised on privacy protection by providing more opportunities to control personal data rather than having legal requirements.

Data controllers under the GDPR collect, store, process, and use personal data only for predefined purposes. However, individuals do not have such restrictions on their data. Individuals can benefit from the flexibility of using this data for the purposes they define. In practice, this can be achieved by allowing individuals to reuse personal data directly and sharing data between services according to their needs and wishes. One of the requirements of MyData is to enable individuals to download data about themselves to their personal PCs or smartphones in a machine-readable format. Furthermore, it is important to keep the latest data accessible through a standardized programming interface (API). Updating data this way allows you to use multiple data in real life without visiting the data provider's website. For example, it is an excellent example of a smart receipt that automatically sends purchase details to a smartphone as soon as you pay for an item.

One of the important outcomes of MyData implementation is the decentralization of the personal data value chain and the centralization of data management for individuals. This will open up opportunities for new entrants and break down the traditional boundaries between sectors and industries. It also opens up the business environment related to personal data so that no single enterprise or institution can realize the monopoly scenario of the future of the data-driven society. It is to prevent the monopolization of personal data. An open personal data ecosystem means avoiding situations where personal data is located in one or a few services or where services based on personal data can be implemented using only one type of technology. For example, the situation in which all personal data is stored in Google or services are used only through Google's technology is the opposite of the direction MyData is aiming for.

The value chain of personal data processing consists of the collection, transmission, management, and utilization of personal data, and the entire value chain has traditionally existed within a single organization. For example, when an individual's account transaction history data is generated in the banking system, the bank processes and transmits the data to provide customers with various services, such as account statements or online banking. Using MyData shifts from a value chain exclusively implemented and controlled by individual banks to an open and distributed value network where companies and institutions at different levels can leverage their personal data. Examples of successful personal data value chain segmentation can be found in the EU's PSD2 (Payment Service Directive 2) and UK banks. After the revised Payment Services Directive was introduced in the EU, the UK CMA (Competition and Markets Authority) decided in 2016 to require major banks to disclose account transaction data to authorized third parties. This decision allowed us to provide services such as personalized financial management based on data provided by several banks. Open banking, introduced in Korea, also began in this context.

MyData by Countries

The start of MyData in the United States is the Smart Disclosure system promoted by the Obama administration. Smart Disclosure is to standardize complex data and make machines into standalone files to help consumers make decisions. The basis for the U.S. smart disclosure system is the

Consumer Privacy Bill of Rights (2011). It allows consumers to choose products and services that suit them by using their own information, such as personal medical records, mobile phone bills, and energy usage history. Information about individuals is disclosed only to the individual himself, and the government and companies that have the data disclose the data to each individual while security is maintained. Chief Technology Officer (CTO) Todd Park conducted the MyData project and provided smart disclosure services by creating buttons for each color (blue, green, red, orange). Blue is medical, green is energy, red is education, and orange is solar. The education sector was originally a red button but was later renamed the 'My Student Data' button. You can download a hospital certificate and an electricity bill as a machine-readable file by playing the button. In particular, blue buttons are well-known in the medical field, including Veterans Affairs. This blue button is located on the home page of public institutions or medical facilities in the medical field. If you click this button after authenticating yourself, you can download your medical history as a file from hospitals connected to the Blue Button service across the United States. You can also install and download a health business created by a private service company. Downloaded files can also be used from another mouth. For example, you can link an insurance app to share medical data and then claim insurance premiums. You can also subscribe to a service that can send SOS signals in an emergency.

In Japan, the MyData industry was introduced as an information bank. We can leave our personal data at the information bank, and the bank sells the data instead and returns the profits to the individual. It can be seen as a kind of personal data trust management business. Just as money is left to an asset manager to increase assets through financial technology, it is a concept that increases assets by using data on its own when personal data is distributed. In Japan, personal data can be collected and utilized without the data subject's consent if the purpose is specified. In the case of consent provided by a third party, the opt-out method can be applied within a certain range. Although it is the first Asian country to receive the GDPR adequacy decision, many areas can be evaluated that regulations on personal data are relatively not strong.

EU has introduced data portability through the GDPR to ensure individual data sovereignty while responding to large platform companies that use data from European residents. After that, data portability was applied to the financial industry through PSD2. The UK also mandated its own open banking policy, the financial sector of MyData. The UK announced the Open Banking Standard 3.0 in 2018 and prepared an open API standard. The UK has implemented open banking for major banks for the first time in the world, and unlike the EU, it has also opened all financial product information. The concept of MyData is also being introduced through smart data policies. Services such as Open banking and Open Finance, Midata in the energy sector, and Dashboard in the pension sector are provided through smart data policies.

The Potential of MyData

Behind MyData is the motivation to create and accelerate change toward a fair, sustainable, and prosperous digital society. Benefits for people and organizations, and ultimately for society as a whole, are the driving forces behind these changes, and nothing will happen without them. Therefore, it is important to find applications in the MyData model that specifically serve people in their daily lives from the start and improve the functions of companies and other organizations. Solutions for the infrastructure and regulation of personal data should create more personal data use while also realizing the benefits of participating groups in the future.

But as with all things, there are potential risks that come with change. Although there will be more demand to use MyData to provide customers with more personalized and special experiences, these customized services have difficulty collecting and analyzing customers' financial and non-financial data to gain insight into them. Personal data protection and management are also required, and efforts to prevent bias in customized algorithms should continue. Nevertheless, the direction of the data finance era that the financial sector is currently aiming for will be hyper-personalized customized services. In particular, the data quality is the most important thing for ultra-personalized customized services. To provide ultra-personalized customized services, you first need basic information such as the customer's age, gender, occupation, and residence, and behavioral data to understand the customer's purchase history, search history, visit page, and interest in the shopping cart. In addition, data to identify emotions and preferences, such as customer product evaluation and reviews, and customer behavior data, such as customer time, place, situation, and purpose, will be needed. In short, financial and non-financial data must be secured in combination. However, financial firms are having difficulties collecting data from non-financial firms, and non-financial firms are demanding more data from financial firms. Therefore, balancing privacy protection with providing hyper-personalized customized services through de-identification measures for collected data is critical for the continued growth of the data economy and the settlement of MyData.

The biggest of such risks is the situation in which MyData does not eventually materialize. Companies of various sizes, public administrative agencies, research institutes, and non-governmental organizations are working to realize MyData. With the advent of attractive concepts and practical implementations, expectations for the future have risen, and more participants, funds, and overall activities have become possible. As the evolution to MyData progresses, there may be unacceptable points in the field, and existing mechanisms may absorb some. Certain technology standards may become widespread; certain companies may secure important market positions; so-called "killer apps" may become de facto standards, or internationally recognized governance models or institutions may emerge. However, in the end, these changes are also expected to be the basis for existing MyData models to evolve into more diverse models. A particular problem to be aware of is the possibility of data leakage in that sensitive personal data is collected and managed. In most countries, the first prerequisite for the success of MyData is 'public trust in privacy protection.' Once trust is broken, it is hard to rebuild. No matter how well we lead the MyData industry, it may be difficult for MyData itself to become a good attempt if cases caused by personal data leakage or misuse become a bad perception.

V.

Artificial Intelligence and MyData

Background

As AI models evolve based on data accumulation, algorithm development, and increased computing power, the AI industry is growing dynamically in various fields. In the future, AI is expected to create innovation and growth opportunities in all areas, such as medical care, education, and distribution, and to contribute significantly to the promotion of benefits and welfare in society. However, changes in data processing methods in AI environments are making it challenging to apply existing privacy principles and standards. The increase in new types of personal data processing, such as data combinations between heterogeneous industries, increases confusion among operators and intensifies anxiety and concern about the possibility of personal data infringement. Therefore, socially acceptable privacy guarantee measures are needed to create a virtuous cycle development ecosystem that can maximize the social benefits of AI. Major countries worldwide are also preparing legal and policy strategies to minimize risks while preoccupying the competitiveness that AI technology and service development can bring. The U.S. NIST established the AIRMF (Artificial Intelligence Risk Management Framework) to manage AI-related risks well and passed the EU's first AI Act.

AI and Personal Data

AI and data are inseparable, and learning data for AI can be identified mainly by dividing it into 'structured data' and 'unstructured data.' First of all, 'structured data' refers to data that can be processed immediately by a computer because the value for a specified variable is stored. Structured data is often composed of rows and columns, and traditional statistical analysis has been conducted on these fixed forms of formative data. In contrast, 'unstructured data' refers to cases where fixed forms such as voice, image, video, and free input text are not taken. The development of artificial intelligence since the

2010s has been centered on using unstructured data because it was believed to be a shortcut to keeping up with human intelligence. Image recognition, voice recognition, and natural language processing are representative examples. Since unstructured data is complex to use immediately for analysis, converting it to a standardized form that a computer can process is common. Discussions on existing personal data protection and de-identification measures focusing on formal data have been discussed. These personal data protection and de-identification measures have been implemented using a privacy model mainly for structured data because privacy models and de-identification measures for unstructured data have not been commercialized. However, this book will try to cover both structured and unstructured data.

Personal data protection has taken a new phase due to generative AI. Existing personal data protection regulations, such as requiring individual consent, are unsuitable for the rapidly changing AI era, and the issue of free use must also be considered. Several companies and the government are stopping using ChatGPT and preparing data protection measures as reports that personal data and sensitive information are leaking through ChatGPT. European countries are investigating whether OpenAI complied with GDPR, warning that huge fines could be imposed. Countries are busy preparing personal data protection measures for this AI era. OpenAI said it would not learn the information of paid customers, which paradoxically means that OpenAI has confessed to using sensitive information as learning data and will continue to use free customers' data in the future. OpenAI is announcing plans to prevent users from learning unwanted data while preventing them from recording conversations with ChatGPT and unveiling new services that can safely protect users' data. Still, it is unlikely to be easy to solve the data protection problem. Business demands to collect and analyze more data to create high-value-added services and to protect personal data and important information are conflicting, and building strong control-oriented regulations to solve this problem will hamper growth. Complex and vast AI requires principle-based regulations, not detailed regulations. It is necessary to protect data while utilizing it freely in all environments safely.

Each country is preparing a principle-oriented data protection and utilization plan suitable for the AI ecosystem. In the case of Korea, the revised 'Personal Information Protection Act,' which takes effect in September 2023, also prepared a plan to allow information subjects to choose to disclose or disclose sensitive information to correct the practice of excessively collecting and utilizing sensitive information. Recently, ChatGPT has become an issue regarding privacy. Even if it is not ChatGPT, information leakage is becoming a severe problem daily. No matter how strong personal data protection regulations and security systems are in place, information leakage cannot be prevented entirely. Attackers cleverly dig into unprepared loopholes, buy disgruntled insiders, or induce users to make mistakes in everyday work environments to leak information. Information leakage accidents due to negligence in management continue even now. Governments worldwide are preparing strong regulations on personal data protection to prevent personal data leakage. Privacy acts do not impose fines on all personal data leakage accidents in general. Penalties are imposed based on the level of compliance with privacy regulations, including the level of information security technology built to prevent hacking and insider leakage, the overall security measures, investment size, and social norms.

In the case of Korea, measures that can be used along with the safe protection of personal data have also been greatly expanded. The right to request the transmission of personal data has been newly established so that the MyData business, which began with the revision of the Data 3 Acts in 2020, can be expanded to all industries. The right to request the transmission of personal data has been established and promoted in some fields, such as public and finance, but it has not been commonly applied to all industries. The right to request the transmission of personal data refers to the right to require a service provider to transmit its personal data to a third party. For example, when you use an SNS service and find other services more attractive, you can move your data to a new service from the existing SNS service provider. Since transmitting to a third institution is difficult, individuals must download and move data individually. In the future, the right to transfer data may also require the intervention of artificial intelligence. In particular, there are studies and attempts to utilize artificial intelligence's learning ability in the case of unstructured data or when standardization of data transmission between heterogeneous industries is not preceded.

Changes in AI Data Processing Method and Privacy Issues

As the focus of AI technology shifts from discriminant AI to generative AI, learning methods are also changing from supervised learning to unsupervised learning. A large amount of non-labeling data is learned, and the types of data that are learned are also multi-modal, from text-oriented to images and images. The use of large-scale Internet-accessible public data, including image and image data related to an unspecified number of people, is increasing. The number of parameters and the amount of data input are growing. The purpose and scope of use are diversifying as pre-learned foundation models are used for various services through fine-tuning or prompt learning. As a result, new privacy issues are emerging. Such is the infringement of an unspecified number of personal data, generating data different from the facts, or exploiting biometric information.

In addition to the possibility of personal data exposure or re-identification, the possibility of profiling or sensitive information generation and processing increases, and large-scale language models generate answers based on language patterns.

Therefore, a systematic approach is needed based on an accurate understanding of AI risks. Due to the lack of systematic understanding of AI's adverse effects and side effects on the basic rights of data subjects, it seems that it will have to identify specific risks. It is necessary to identify and evaluate risks based on the probability and severity of problems caused by AI and to set regulatory directions by type and sensitivity differentially. In addition, it is necessary to resolve uncertainties about data processing in AI development and service. Since it is essential to use public information to secure and learn high-quality data when developing AI, it is necessary to clarify measures for each data processing stage and the level and method of pseudonym processing of unstructured data beyond simple identification. Of course, as AI services without borders expand, individual countries' regulations alone have limitations, and it is difficult for companies to comply with different norms. Still, international discussions will eventually be held based on sharing various trends, such as policies and investigations on AI.

Biometric information generated through specific technical means is sensitive information and should be used for AI only if there is separate consent or legal basis. Alternative means should be prepared when processing biometric information, the original information should be kept separately from other personal data, and biometric information protection measures such as encryption should be implemented when storing. However, biometric information processed to find out characteristics (age, gender, etc.) about an individual, not for authentication and identification purposes, will be able to be processed as general personal data, not sensitive information in most countries. For example, classifying users by estimating age and gender through facial recognition can be regarded as general personal data processing, not sensitive information. On the other hand, extracting, collecting, or generating and processing biometric information from information disclosed through crawling, etc., needs to be strictly limited. Of course, in this case, extracting and processing biometric information to identify individual characteristics may be the same as general personal data, but processing for other purposes may constitute sensitive information processing. So, we should note that this processing of personal data using artificial intelligence is sensitive and can produce unwanted results.

Data Processing Criteria and Protection Measures for the Artificial Intelligence Stage

In the artificial intelligence model and service planning stage, it is necessary to reflect the design principles centered on personal data protection. When developing AI models and services, the Privacy by Design and Default principle must be applied, considering users' privacy from the planning stage to the entire personal data processing life cycle. In particular, since AI processes a large amount of data in a highly complex manner, it is necessary to remove risk factors by reviewing the legality of personal data collection and the possibility of privacy infringement in the modeling, learning, and operation process in advance. So, businesses that want to develop and provide AI models and services should evaluate risks in advance by considering the data learned by the AI and the characteristics of the services provided and prepare strategies to reduce and minimize risks according to the evaluation results.

In the data collection stage, clear processing standards and application case guidelines are needed for the scope of collection and use of personal data necessary for AI development. Personal data legally collected according to contract signing, implementation, and compliance with laws is expected to be used for AI development and service within the scope of collection purpose, and in most countries, AI development and service are predictable and do not unfairly infringe on the interests of data subjects. Of course, the criteria for that judgment are expected to vary from country to country. In addition, if consent is obtained for collecting and using personal data for AI development and service purposes as an expression of the data subject's voluntary consent, it can be used within the scope of consent. The problem is when the disclosed information is used as learning data. When developing a large-scale language model, it may be necessary to use public data, as it can also be processed according to personal data processing standards. Still, contract signing, implementation relationships, or consent are often not established. Nevertheless, as a result of the sentence of profit on the processing of disclosed information, if the profit from processing is recognized to be greater than the profit from preventing it, or if the legitimate profit clearly takes precedence over the right of the data subject. In the data learning stage, it is necessary to actively utilize Privacy Enhancing Technology, suitable for personal data utilization and processing. Technologies such as de-identification processing technology, synthetic data, differential privacy, homomorphic encryption, and federated learning are increasingly important with AI and big data development. In the AI service stage, data subjects should clearly know how their personal data is collected and processed during AI development and service, and operators should notify the purpose and method of processing data used in AI model development and service based on personal data processing policy. It is necessary to actively refer to expandable AI, which has been studied a lot recently, for a specific scope of disclosure, contents, and notification measures. In addition to enhancing transparency, it is necessary to prepare procedures to prevent unfair infringement of the data subject's rights when using public information and unspecified multiple photographed videos and to guarantee the rights of the data subject so that the AI service can easily understand and exercise their rights.

Language Processing Artificial Intelligence and Learning Data

Processing or understanding natural language is a major application field of AI. Textual emotion analysis analyzes reviews of products and services left by customers through AI and automatically checks whether online posts or comments comply with the terms of use through AI. Media companies write standardized articles using AI in writing in some areas. AI is active for educational purposes, such as evaluating writing test results. In addition, several AIs that perform specific tasks through conversations with human beings, such as artificial intelligence assistants and artificial intelligence speakers, have been announced, which have the potential to be used in various forms in the future. In particular, it is noteworthy that the large-scale language model has recently attracted attention in natural language processing. A language model is an AI model based mainly on a statistical methodology that calculates the probability distribution of a sentence. Therefore, building a highly accurate language model is becoming a deep task in natural language processing. There are many methodologies for building language models using AI, but recently, several large-scale language models based on the Transformer language model were released by Google. OpenAI's GPT-3 model unveiled in 2020 is representative and stimulated by its performance.

The development of large-scale language models is being competitively carried out. Such artificial intelligence often has hundreds of billions of parameters. AI based on this language model is now difficult to distinguish from human writing. It not only fully complies with grammar, but its content and sentence composition are also natural and logical. Such a large-scale AI model requires a large amount of learning data. In general, hundreds of gigabytes or more of data are used for learning by collecting and refining tens of terabytes or more of data published on the web. Various products, services, movies released online, reviews of music, posts, and comments from multiple communities, and social network service conversations are also used as learning data, with a capacity of several terabytes. The legal regulations on using such disclosed personal data for AI learning purposes have not been clarified clearly.

On the other hand, there may be cases in which personal data is included in the text data collected by the consent from the data object. Examples include conversations with customers, e-mails, and call center calls held by companies. In addition, there are cases where customers provide conversation data while using their mobile application. As for the text data collected with the data subject's consent, it is unclear whether it is within the scope of the data subject's consent to learn it from AI, which can become an issue these days. In addition, even if confirmed, it can be assumed that the data must be used for artificial intelligence learning by taking de-identifying measures to protect the data subject's privacy.

De-identification of Personal Data for Learning and Re-identification by AI

Personal data included in the text often takes a predetermined format. A considerable number of personal identification information can be found through certain rules. Most names, phone numbers, account numbers, credit card numbers, addresses, e-mails, IP addresses, MAC addresses, and home page URLs can be found according to fixed positions. In this case, the rule can be formalized using regular expression, and the corresponding personal identification information can be found through the rule. The legal regulations on using such personal data for AI learning purposes have not been clarified. On the other hand, there may be cases in which personal data is included in the text data collected by the consent from the data object. Examples include conversations with customers, e-mails, and call center calls held by companies. AI for text de-identification is already commercialized, and it is also possible to re-learn and utilize the existing AI model. Commercial de-identification measures can be applied immediately, but the accuracy may not be high depending on the specific purpose. It may be more accurate for individual service providers to develop or re-learn on their own. Still, there will be additional burdens, such as building additional learning data.

When personal data is deleted from the original text, the meaning and structure of the sentence may change, which may cause problems. In other words, there is a risk of results that are not appropriate for AI learning. To prevent this problem, personal data may be replaced with other texts of similar form to the extent that the structure of the sentence is maintained. It's a sort of pseudonymization. However, it is also difficult to establish detailed standards for replacing such personal identification information in practice. For example, information about celebrities is part of common sense, so AI may need to learn it. Therefore, in the case of matters concerning the person registered in the biographical dictionary, it is also possible to consider a plan not to replace it intentionally. However, it is not easy to determine whether the person mentioned in the sentence in the learning data is a public person, and it is still a very difficult task to automatically distinguish whether the person is a celebrity or a person with the same name. On the other hand, even if text de-identification technology is used, there is a risk that personal identification information cannot be deleted entirely and replaced. As a result, the usefulness of data may be reduced.

When the text collected with the data subject's consent is used for AI learning, it is necessary to interpret the scope of the data subject's consent. In this regard, there has been a notable incident in Korea. A Korean startup collected messenger conversation data from users and stated in its privacy policy that the conversation data provided by the user could be used for "new service development." Considering the context of the conversation, the chatbot operates by selecting and answering the most appropriate response from among conversation sentences pre-stored in the response database, and the above data was used to learn the chatbot's ability to select the appropriate response. In several stages, the response database included more than 100 million user conversation sentences, which were de-identified by excluding names, addresses, numbers, etc. In this case, the issue was whether the user's consent to use conversation data for 'developing a new service' could be interpreted as consent to use chatbot AI to learn. In conclusion, the case concluded that it was difficult to say that users expected and agreed to use their personal data to develop and operate new services completely different from existing services just because 'new service development' was specified. In this case, the PIPC in Korea used the main criteria for interpreting the scope of consent, such as agreement with the user's intention, the expected possibility of the user, and the possibility of unexpected damage. In practice, if personal data is included in the published text, the use of the personal data must be 'objectively recognized as having the consent of the data subject.' It can only be used without the new consent of the data subject. If this is unclear, de-identification processing may be used for statistical and scientific research and preservation of records in the public interest to eliminate the possibility of identification.

Data released on the Internet has been widely used for the learning of AI in the field of computer image processing. Still, the learning image and image data often contain information that can identify individuals. As in the case of text data, whether the data can be used without the data subject's consent is a legal issue. For example, AI for face detection will not pose a significant risk to privacy. Still, it can be problematic if features are extracted from the face and learned to detect a specific face in the video to find the criminal. In the GDPR, this can be profiling, so notification and consent to profiling are required. However, the problem does not stop here; there is much room for controversy over the use of photos or videos posted by data subjects through

SNS. Of course, further research and social consensus should be made in the future, but it isn't easy to conclude.

As for AI, it is challenging to identify algorithm errors. For example, in the case of deep learning, unlike general programs, it is difficult to determine what data caused the error during development because the clear basis for the derived results is often unknown. The disclosed personal data is frequently used for AI learning. In particular, learning data required for AI learning is usually collected through web crawling techniques, and as a result, personal data is often included in the collected data. Issues raised in this regard under privacy acts include whether the disclosed personal data is available and, if possible, how far the allowable range is. In conclusion, many theories can use the disclosed personal data without the data subject's consent. Still, the scope and conditions of its acceptance vary from country to country and case to case.

It has been thought that the privacy risk is insignificant if there is only a learned AI model after learning is completed. Recently, however, a model inversion attack scenario that can attack reversely and extract personal data from learned AI has also been studied. For example, it is possible to recreate the original image used as learning data quite similar to the actual image by utilizing machine learning output. In the case of AI that creates sentences, such as chatbots, there is a risk of regenerating personal data included in learning data, which happened a few years ago in Korea. There was considerable controversy then, and the company had to pay hefty fines. Most of these results were not initially expected by the data subject. As a result, it is unlikely to be suitable for personal data-related compliance, whether it is using public personal data or with consent. Because of this risk, personally identifiable information should be removed as much as possible when constructing AI learning data, and de-identification measures should be taken. Of course, it is unclear how realistic the possibility of an attack is, but there is always a risk of regenerating personal data contained in learning data based on AI.

AI-based MyData for Data Economy

MyData is expected to be a model example of the data economy using AI. Collecting and analyzing data in various fields without restrictions on data movement can provide more granular personalized services, create new services that have never existed, and grow the market. Consumers refuse to use their data at their disposal but are active in data sharing if they can provide their information and use better services. However, no consumer can tolerate using it without data protection. It is no exaggeration to say that the success of the MyData business lies in 'security.' MyData service without sufficient security will fail eventually.

Pseudonymized data can be used without consent—in

most countries–but there is minimal experience in using data after pseudonymization. In Korea, as of 2023, the public sector accounted for 30.1%, and the private sector accounted for only 5.2% of the experience of pseudonymization. The Korean government has worked on various measures to revitalize the MyData industry. As the biggest obstacle to MyData is privacy, it is taking a literacy improvement policy for MyData by recognizing that MyData can actively control its data by securing data sovereignty of the data subject. Based on key institutional and technical infrastructure, strategies are established to revitalize the MyData ecosystem, such as services and incentives, and to secure reliability at each stage of use of MyData. In addition, detailed standards for the legal system will be established to ensure the exercise of the right to data portability in which the public actively manages and controls personal data, and a platform will be established to technically support the process of using MyData. PIPC in Korea plans to promote standardization of common data format and transmission standards so that data can move between heterogeneous industries and sectors without blockage and expand people's data benefits by strengthening the promotion of MyData leading services.

However, for the time being, one of the tasks that must be established to revitalize MyData can be data combination in heterogeneous industries, and securing pseudonymized data is needed to utilize information from subjects who disagree. To introduce one of the selected cases for PoC service is to develop a predictive model for high-risk vulnerable elderly by analyzing the factors that cause older people to enter vulnerable groups such as low-income, welfare, care, and medical benefits recipients. To develop this prediction model, it combines the basic pension data of the Social Security Information Service, insurance, medical treatment, and work history of the National Health Insurance Service. In addition, many scenarios closely related to people's lives and have high policy ripple effects were selected, such as calculating and revitalizing the living population to support demographic areas and analyzing the impact of work-family balance policies on female workers' economic activities and childbirth.

The PIPC in Korea provides an appropriate solution for pseudonymization through a comprehensive support platform for the use of pseudonymized data and supports the entire process of combining pseudonym data, such as reviewing the appropriateness of confirming that pseudonym processing has become safe. As such, MyData can be an example of a data economy model using AI, and if there is a government guide in the early stages, it can help spread the industry.

Another world, Metaverse

The term metaverse comes from a combination of meta and universe in the 1992 science fiction novel Snow Crash. In fact, for Generation X, who first experienced a network called the Internet, which did not exist in the past, the metaverse tends not to feel so new. Still, it is not too much to understand as a virtual world network focused on social connectivity. These days, the metaverse often includes augmented reality, life logging, and virtual reality, but we can say that there is a paper difference from the online MMORPG (Massively Multiplayer Online Role-Playing Game).

But if all human daily life is done in a digital world such as a metaverse, not a real world, which world is real? These questions have been consistently asked, regardless of age, age, and age, even if we do not mention Zhuangzi's butterfly dream. In the movie Matrix, Morpheus told Neo, "What is real? How do you define 'real'?" is also a conclusion in the same context of existence. Aside from complex philosophical thinking, we are already becoming human beings who exist in many kinds of worlds, not in one world. We don't know how we will live in the metaverse. Still, assuming that in the future, metaverse-related technologies will gradually develop and expand into another "real" world as a platform, it would be reasonable to say that we have all the problems in the world we live in now. When referring to the metaverse, in addition to the technologies of existing social/game platforms, the AR, VR, and de-centralization technologies (blockchain, NFT, web 3.0) are often mentioned as elemental technologies. However, to establish itself as a new world and space for humanity to lead daily life rather than technology, the social community must survive, such as law, institution, and social and ethical concerns.

Metaverse is a digital world. Since individuals also exist as digital data, not as physical, the metaverse has the characteristics of digital data as it is, and there are various problems to consider due to the infinite replication and reusable characteristics of digital data. Copyright, publishing rights, trademark rights, design rights, and ownership (for items within the metaverse) have recently been highlighted. If we move the world we live into the virtual world and reproduce it as it is, it will likely constitute an infringement of these rights. Of course, buildings and exhibits always displayed in open places are subject to copyright exceptions. Still, detailed discussions such as copyright fees will soon be needed, considering that many performances and exhibitions are held in the metaverse due to COVID-19. Therefore, legal and technical preparations must be made to support this. In particular, blockchain-based NFTs (Non-Fungible Tokens) are already being actively commercialized for item ownership. Since all objects in the metaverse are basically digital data, ownership is not easy to recognize, but NFT enables at least proof of rights for particular items. While blockchain-based cryptocurrency is basically FT (Fungible

Token), ownership within the metaverse will be protected by recognizing legal validity by relying on blockchain technology for the same effect as registration for individual items.

Considerations for Metaverse & AI

As people spend more time in the metaverse, ethical and social problems arising from real communities are more likely to be reflected as they are. Unlike the real world, since the metaverse presupposes a platform, personal data-related matters can be the biggest problem. It seems more difficult to legislate on personal data protection within the metaverse and to exercise practical rights. There are many teenagers in the main user base of the Metaverse platform that is already in service, so more attention is needed. Since platform companies collect all information on activities within the metaverse and biometric information can be collected through AR/VR, legal requirements for consent, collection, and utilization of data subjects should be strictly applied. However, since individuals in the metaverse are game characters or avatars, it is unclear whether collecting and using those data violates privacy.

Due to the sexual harassment incident within Horizon Worlds by Meta, the addition of a function to prevent unwanted interactions by adding a 4-foot distance between avatars by setting an invisible space called "Personal Boundary" around the avatars. The VR device (Meta Quest2) used to use Horizon Worlds can recognize the user's hand, interact with the hand, and voice chat within VR, so ethical and social problems in the real world can occur equally. Current VR devices are audiovisual-oriented reproduction, but if extended to the rest of the senses, including touch, it is difficult to dismiss the interaction between avatars as a virtual digital world. Meta's Reality Labs is already developing a haptic glove, and VR Electronics is also developing the Tesla Suit, which allows you to feel wind, heat, pain, and water.

Metaverse companies are expected to collect personal data to a previously unseen extent for identification, advertising, and user tracking through various channels, including AR/VR wearable devices, microphones, heart/breathing monitors, and user interactions. Meta's Morea said the next VR device, Project Cambria, can mirror users' facial and eye movements. The Extended Reality Safety Initiative (XRSI) asked to share details about data collection plans in the metaverse, but Meta Reality Labs refused to disclose the schema. Microsoft, which is creating its virtual reality software, has also avoided answering whether or not to collect and share users' biometric data and plans using the technology. This shows the danger that if the metaverse is operated incorrectly, it can easily lead to a world without privacy. There is a possibility of collecting much more personal data, from conscious physical activities such as eye movements to unconscious blinking, and such biometric data can further be individually identifiable data. In fact, not only are biometric data such as fingerprints and iris used for user authentication, but also a considerable level of user identification can be used for gait, signature, and gesture patterns, so attention should be paid to collecting and utilizing biometric data. Biometric data used for user authentication, such as fingerprints and iris, should not leave the user. Due to the nature of biometric data, it is impossible to change it when leaked, so neither Samsung Electronics nor Apple leave mobile devices for fingerprint and iris data. One of the biggest risks is that these security and privacy principles have not yet been established in the metaverse world.

However, the industry cannot develop only when establishing security and privacy principles is always prioritized. For AI and metaverse, which have complex data utilization ranges and methods and a fast pace of change, regulations centered on "principles" rather than detailed "regulations" are needed. It is necessary to design personal data regulations for communication and cooperation between countries, the government, and the private sector. To this end, it is necessary to establish a system in which the government, AI developers, and operators participate and discuss together from the design stage of a disciplinary system. For example, establishing an AI risk assessment model or a framework such as GDPR's adequacy decision can be a good approach.

VI.

Approaches to Realize MyData

Background

Many fields, such as finance, medical care, health, and administration, are paying attention to MyData. New attempts to use personal data have also begun in the general consumption industry. Still, in the end, new innovative services are possible only when heterogeneous data between industries are combined. The opposite benefit is privacy. A new approach is needed to solve this problem fundamentally. Recently, in the field of ICT, big data and artificial intelligence have been constantly studied in the industry and academia, and services that can be applied in real life have been developed. However, they are still talking about garbage in and garbage out. In the end, it is concluded that securing high-quality, real-time personal data with a lot of information is the key to innovative services. Much information about individuals eventually means combined data, and the MyData industry seems to be jumping into this field to get personal insights and businesses by combining personal data with personal consent.

However, these business-oriented MyData services are inevitably very likely to be a problem in terms of security and privacy. There is no need to explain that the security risk increases when data is concentrated in one place, and the level of profiling increases when personal data is combined, so the privacy risk increases. It is difficult to say that the risk of concentrated data and combined personal data is risk hedging simply by complying with the regulations related to network separation in the Electronic Financial Transactions Act and the Credit Information Act. If non-financial data are not just financial data but also combined data, hacking attractiveness will inevitably increase. This may also be ironic, as the maxim in the investment sector not to put eggs in one basket can also be applied to the global MyData industry.

When discussing the privacy risks of using personal data, the part often mentioned is using pseudonym data. The EU's GDPR also recognizes pseudonymization as an important protective measure for minimizing personal data and as an additional protective measure that can help determine the compatibility of the original and new purposes of the processing. However, pseudonym data with identifiable links compromises identification and anonymous data so that it can meet requirements such as data combination. Still, it isn't easy to use beyond statistical purposes that cannot be personally identified unless the user agrees to the combination. Not a few MyData operators seem to be missing this part. In the current structure and approach, protecting privacy and utilizing data simultaneously is difficult.

The movement of global companies to protect personal data and the current server-based MyData services seem to be heading in the opposite direction. Due to privacy issues, federated learning models and deep learning on-device AI are attracting attention in machine learning, and focusing personal data on servers of MyData operators is not desirable in terms of security and privacy protection and does not fit global trends. Therefore, an innovative approach is needed to prioritize user privacy and increase corporate utilization. It is time to actively consider a new MyData service model, such as the form in which individuals combine, analyze, and process their data on users' mobile devices, as a disruptive innovation approach.

PDS (Personal Data Stores)

The European Strategy for Data, published by the EU Commission in February 2020, repeatedly mentioned PDS but did not precisely define it. Instead, there have been many thoughts and attempts on PDS in the private sector, and as a result, projects such as open PDS/SA (Open Personal Data Store with Safe Answer) were created. It is a form in which an application uses, analyzes, and answers the user's personal data using the SafeAnswers framework. Open PDS/SA provides code rather than data, turning the difficult data anonymization problem into an easier security problem. SafeAnswers uses two separate layers to aggregate user data. Sensitive data processing takes place within the user's PDS. It allows data to be aggregated anonymously between users without sharing sensitive data with intermediate objects. General calculations on user data are performed in a secure PDS environment under the user's control, and the user does not need to hand over the data to receive the service. Only the answers and summary data required by the app are sent outside the user's PDS. For example, instead of exporting GPS data, the app may be sufficiently serviceable with information about whether the user is active or in what general geographic area he is currently in. Instead of sending GPS coordinates to the company's server, the calculation is performed within the user's PDS by the corresponding Q&A module, and this attempt alone can achieve a high level of privacy.

SOLID is one of the similar projects. It is a decentralized web project led by Tim Berners-Lee, the father of the web, and being developed in collaboration with MIT (the Massachusetts Institute of Technology). This project aims to create true data ownership and privacy improvements by fundamentally changing how web applications operate today. Solid's goal is to ensure users' personal data is completely protected. What gives you control includes access control and storage control. Tim Berners-Lee also established a separate corporation to build a commercial ecosystem through Solid.

Recently, industries and academia in the ICT field have constantly studied big data and artificial intelligence and developed services that can be applied in real life. However, they still discuss GIGO (Garbage In, Garbage Out). In the end, it is concluded that securing high-quality, real-time personal data with a lot of information is the key to innovative services. A lot of information about individuals eventually means combined data, and the MyData industry seems to be jumping into this field to get personal insights and businesses through the combination of personal data.

Though MyData Global Association supports the MyData operator model, operator-centric MyData services inevitably have a high potential for problems in terms of security and privacy. There is no need to explain that security risks increase when data is concentrated in one place, and privacy risks increase as the level of profiling increases when personal data is combined. It is difficult to say that the risk of concentrated and combined personal data can be resolved simply by complying with the regulations. If it is not just financial data but non-financial data combined, hacking attractiveness will inevitably increase. This may also be ironic, as the maxim in the investment sector not to put eggs in one basket can also be applied to the situation of the MyData industry. This kind of compliance huddle is not easy to overcome due to the characteristic that privacy is a human right and personality right, so we should approach with a technical solution, which the technical perspective and a new approach to PDS will be covered later in this book.

Cookies

The most common use of individual behavior data will probably be cookies. A cookie is a small file of historical information installed on an Internet user's computer or other device through the user's web browser when an Internet user visits a website. Usually, it stores a history of visiting the site, usage records (e.g., search terms, login status), etc. The word originated from a magic cookie that Unix programmers used to mean packets of data received and returned unaltered by programs, but it is not very new to the public. This is because more websites ask, "Do you want to allow cookies?" Before these questions were asked, cookies had been deeply involved in our Internet use. Many people were surprised when I saw the word I searched, the advertising banner related to the site I visited. However, it has become a somewhat familiar user experience these days. However, in 2020, the term 'Cookieless' began to appear frequently. What function did cookies have, and how should we prepare for an era without cookies?

Cookies contain data about users. It is used to identify users, and cookies are utilized in many applications, from maintaining login sessions to delivering advertisements through context targeting. Of course, cookies are created on a browser basis, not on a user basis. Still, since most users use the same browser on their devices to navigate the web, they will likely identify individuals just by chasing browser records. Therefore, cookies can also be said to be files in which personal data is recorded. Cookies can be divided mainly into first-party cookies and third-party cookies. A first-party cookie is a cookie file issued directly by a website you visit. For example, when a user accesses 'google.com,' it is a first-party cookie that google.com issues about the login history. If you check options such as "Keep logged in" on the login screen, you save Google login status, close the window, and then log back in, and it still stays logged in, which utilizes cookies issued by Google. Third-party cookies are issued by a website other than the website visited by the user, and this is usually the case when other sites use cookies issued by the advertising server.

Cookies are often regarded as personal data because personal data is usually stored in cookies or cookies and used to provide customized services. Still, technically, simple information generated while using the service is complex to see as personal data. Nevertheless, when an individual is identified in connection with the individual's identification information, it corresponds to personal data, and it becomes personal data when the user is easily identified in a mobile environment or when it is possible to identify an individual by combining or combining several cookie contents. Usually, when discussing the topic of "blocking cookies," most of them refer to third-party cookies. This is because first-party cookies are used to identify users on websites visited by users, so if they are blocked, their usability is very low. On the other hand, since most third-party cookies are used for online advertisements, blocking or erasing them is not inconvenient. If I have to find something uncomfortable, I may see advertisements that are less relevant to me. However, when these third-party cookies are personalized, privacy protection is needed, and this is why social demand for blocking third-party cookies that increase identification and profiling levels for customer targeting is rising.

Protecting and Utilizing Personal Data
Without Advertising Identifiers and Cookies

Additionally, Apple's App Tracking Transparency (ATT) policy is frequently mentioned when discussing cookieless. Identifier for Advertiser (IDFA), an advertising identifier that allows mobile advertising media to track users, is available only when users allow it. Google also announced the Android 12 version, announcing that a separate permission should be added to collect GAID (Google Advertising ID), commonly called ADID (Advertising ID). In particular, it was announced that the same policy will be applied to Android 12 or lower devices starting in April 2022. Tracking using advertising identifiers such as GAID/IDFA in a mobile environment has a difference in influence in terms of privacy from the PC environment. Statistically, more cookies are maintained by mobile users than by PCs, and unlike PCs, in the mobile environment, the advertisement identifier itself is bound to be a practical personal identifier.

Cookies and advertising identifiers offer advertisers, marketers, and site owners many benefits in terms of utilization, but because of the enormous social costs of security and privacy protection, policies related to cookieless policies and advertising identifiers such as Google and Apple have been announced, and consumer movements have been blocked. What should we prepare, and how should we prepare in an era when the use of cookies and advertising identifiers is decreasing?

Companies first use less identifiable alternative means. Both Google's FLoC and Topic were proposed in the privacy sandbox as alternatives to lowering identification. They are grouping strategies that diminish individual identification by grouping them by interests and topics. Contextual targeting, which displays advertisements in the appropriate context, such as posting sneakers advertisements in health-related articles, is also a suitable method. The use of on-device solutions is also a good approach. This is because instead of marketing to individuals based on data from mobile devices, mobile devices allow access to only as much information as necessary; users can target based on proven user activities while maintaining anonymity. More and more people have recently been using such on-device data in the MyData area.

Cookie-free environments are coming in fashion for privacy, but there is also a paradoxical aspect: The value of protection increases when personal data is used. The cookie-free future promises many benefits for privacy and security. Still, it will also bring challenging times for many companies that already use cookies to market and target customers, reducing the value of privacy. Nevertheless, the issue of personal data protection is still important because personal data is not just data but also human rights or personality rights. It will not be easy for ordinary users to understand this situation. Still, it is also an obstacle to overcome, and meaningful social changes can be expected when overcome. It's as if, at some point, we've learned over a long period that it's better not to check 'optional' instead of unconditionally checking all the checkboxes, and there's more spam than before.

If third-party cookies are discontinued, privacy protection will be strengthened overall, while the user experience may be partially regressed. It may be annoying for the homepage to keep asking about the cookie settings or for the mobile app to ask, "Do you want to allow tracking?" The problem is that this annoyance tends to be relatively insensitive to privacy and security. We have already had the experience of agreeing without asking and without reading the same personal data usage. The intention was to inform the user of the details of the use of personal data, but it resulted in difficulty in adequately recognizing the use of personal data.

According to Accenture, 83% of consumers are willing to share their data for a more personalized experience. The problem is that in the cookieless era, first-party or zero-party data (customer data provided under the premise that customers agree to collect information) will be more emphasized than third-party, and more personal data may be gathered at this point. To overcome this paradox, it is important to improve privacy literacy so that individuals can exercise the right to self-determination of personal data. The cookie-less policy is necessary for protecting personal data.

Still, it is not a sufficient condition, and activities such as carefully watching various consent related to personal data are still up to consumers.

Industry Resistance against Cookieless

These new changes, which make it difficult to use third-party cookies, are suitable for personal data protection but have the disadvantage of making digital marketing difficult. Google has introduced the Federated Learning of Cohorts (FLoC) API through a project called 'Privacy Sandbox' instead of using cookies from browsers and began testing in the form of target marketing based on the same interest group but hiding personal identifiers. Browsers with FLoC enabled can collect information about user behavior and group different users into groups according to their common interests. The browser then shares its collective ID with websites and advertisers. This proposal did not go down well with privacy advocates and browsers. The Electronic Frontier Foundation pushed back, calling it "Google's FLoC Is a Terrible Idea," and argued that discrimination based on search history could be perpetuated. All major browsers except Google Chrome have not adopted this technology. In January 2022, Google officially shut down FLoC's services and proposed a replacement, Topic API.

Apple also reportedly blocked third-party cookies

through ITP (Intelligent Tracking Prevention) and is changing user interfaces in many scenarios to obtain user approval when collecting search and visit records. The movement of international companies to protect personal data and the current server-based MyData services seem to be heading in the opposite direction. Due to privacy issues, Federated Learning models are drawing attention in the field of machine learning and On-Device AI in the area of deep learning. Focusing personal data on servers of MyData providers is not desirable regarding security and privacy protection and does not fit global trends. Therefore, an innovative approach must consider user privacy first but increase corporate utilization. It is time to actively consider a new MyData service model, such as how individuals combine, analyze, and process their own data on users' mobile devices.

Other Considerations and Regulations

Interest in using personal data is increasing worldwide, and all kinds of services are evolving and customized based on personal data. For personalized services, understanding individuals is essential. For this, a three-dimensional analysis of each individual based on personal data distributed and managed by various companies and institutions is necessary. The problem is that it is difficult to manage and control personal data fully in this process, and individuals are excluded from the benefits of using personal data.

Beyond existing MyData in the financial, public, and medical sectors, public-private data must be used to provide integrated, preemptive, customized public services. At this point, transmission and utilization of MyData between heterogeneous industries became essential. In Korea, which has increased in all industries in a short time, the government is used to intervening for the industry to grow. However, there are many points to consider: whether what was right in the past will continue to be correct and valid in the future and whether it will be good in other countries. In particular, regardless of whether having a so-called competent ministry in terms of MyData between heterogeneous industries is desirable, it is questionable whether it will be practically effective in regulation or somewhat hinder industrial development.

The definition of a term is always important. To share a mutually promised understanding, one of the first things to do to avoid errors in all kinds of communication is the definition and unification of intent. Therefore, regulation of terms is essential, and it is necessary to be careful, especially when new industries are introduced. However, it was a surprising idea beyond the monopoly of terms to prohibit the use of the term "MyData" in trade names or names except for those permitted by the FSC (Financial Services Commission) in Korea. This is still valid as of November 2023. Some institutions use public MyData from the Ministry of the

Interior and Safety and medical MyData from the Ministry of Health and Welfare, and the Personal Information Protection Committee is now introducing MyData in all fields. Reflecting regulations that are not practically possible into law will only confuse society beyond making the law useless. As Goldman Sachs Chairman Lloyd Craig Blankfein frequently says, "Our company is an IT company, not finance," it can be proof that financial companies are no longer dealing with money but are becoming data-handling companies. However, excessive scalability can have consequences that interfere with other industries.

Regulation by Acts

In Korea, one of the representative regulations of financial MyData is the prohibition of scraping technology. Starting January 5, 2022, scraping has been completely banned for MyData, and financial MyData operators have been providing financial MyData services to all users only through API methods. It claimed to strengthen data protection and security, inquire more data quickly and conveniently, and connect with more diverse information providers. Still, it is a regulation focusing more on entry barriers than opening financial data. This was a not-so-founded MyData policy, and the financial institutions still use scraping technology to file various documents. Companies such as Mint and Yodlee, which provide leading financial MyData services in the United States, still use scraping. Data protection and security can be supplemented. If security and privacy requirements are met, technology is a matter of choice depending on the pros and cons, not subject to regulation.

Scraping is just a data extraction technology. It is only a technique for extracting data from human-readable digital outputs. In 2018, the FSC claimed that it was an inevitable measure to protect important personal financial data and that it was to prevent data collection by receiving customer authentication information. Server-based scraping can be criticized as such, but banning scraping itself by crushing client-based scraping is also beyond the principle of administrative regulation. This is because client-based scraping does not receive customer authentication information or access by proxy. The FSC's original purpose is not to block technology but to prevent the arbitrary distribution of financial data without the FSC's permission, and it seems practically difficult to achieve the goal if the method of reaching it is a method that restricts technology.

my:D service operated by SNPLab Inc. is a representative example of a head-on violation of these regulations. In the case of my:D service, financial data was thoroughly scraped only on a person's smartphone without storing it on a server (now it is using lookup API). Ultimately, the FSS (Financial Supervisory Service) had no choice but to judge that my:D service was not subject to permission from its credit information management business even though it showed and managed personal financial data on the smartphone. Achieving administrative purposes by regulating technology is a product of valid experience in the regulatory industry and may still be somewhat valid. But the world is already changing. In an era when even virtual currency is now inevitably accepted by institutions, achieving the original goal of regulation in a way that limits technology is complex.

Regulation by Omissions

This can be unique in Korea; 'the right to data portability' mentioned in GDPR encompasses the right to receive personal data and the right to transmit it. It is presumed that the Financial Services Commission wanted to introduce only the right to transmit under the name of transmission request. Instead of expressing the right to data portability, the phrase 'the right to request transmission' can also be interpreted as a regulation using terms. This is because if the right to move implies an active power performed by an individual, there is much room for interpretation that the transmission request is made only when I make a request, and I can only request instead of having it. Of course, The Credit Information Act stipulates the right to request the transmission of credit information subjects themselves, and simply by looking at the phrase, there seems to be no problem because the information subject is included in the transmission target. However, the GDPR divides the right to move data and transmit data into equal levels because they are equally important, and the first mentioned is probably a more basic right. The MyData development policy published by the Korean government analyzed that the MyData business develops in the order of inquiry, storage, and transmission requests according to the degree of data opening and utilization. However, to store financial MyData, individuals can only save the data as a PDF file through an authentication process and a complex consent process through the Korea Credit Information Service's MyPDS application. Even an ice cream has a variety of distribution channels. Distribution channels must be diverse so that the utility felt by users can be improved. The more varied the distribution channels of data, the more active the use of data will be in all industries. Individual financial data can be used in other industrial areas, and diversifying distribution channels is the way to revitalize the industry.

There are reasons for all regulations, but regulations used for territorial disputes between ministries, hinder the development of other industries, or deviate from their original purpose (human-centric in the case of MyData) need to be boldly reviewed from the ground up. In addition, there is much room for new regulations not to improve existing regulations that do not conform to industrial development. To lay the foundation for private-led innovative growth, it is necessary to innovate existing regulations, not do nothing.

Compensation is the Key

Individuals are data subjects, but IT companies monopolize the value generated by data. To overcome this problem, Korea's Gyeonggi Province conducted the world's first "data dividend" to residents on a trial basis, as mentioned in the previous chapter. It is a method of processing and analyzing local currency usage data written by residents as de-identified big data and selling it to companies and others to directly return it to residents who contributed to the data creation. It was a type of dividend paid by dividing the profits from big data sales in local currency by the number of residents who contributed. Although the first start was very insignificant, 120 Korean Won per person, it is significant in direct compensation for data generated by individuals. This is because appropriate compensation criteria for data utilization have never been set so far, and in the end, compensation has never been paid. If data is also a kind of product, the price can be determined by the cost, competitor, and customer, referred to as the 3C pricing model. However, personal data requires an entirely different approach. This is because personal data is often a by-product of individuals' use of the service, so individuals do not cost to generate personal data, and each individual does not compete with each other as the only subject, so the price is mainly determined by corporate customers who want to use the personal data. Since this part is not a technical problem, it is an area that requires a humanities approach. In the future, lengthy discussions will occur in academia, institutions, or social groups. Still, most are skeptical about whether it is a matter that can be concluded. Nevertheless, to be a sustainable ecosystem, some form of reward must be given to individuals who are data subjects. Despite numerous controversies, it cannot be denied that the cryptocurrency ecosystem was expandable because the reward system for participants was eventually well-established, which will be the same for the MyData ecosystem.

VII.

Technologies for MyData Sovereignty

Background

This chapter will talk about the various technologies used to implement data sovereignty. As mentioned earlier, the fundamental reason for guaranteeing 'the right to self-determination of personal data,' which is the basic idea of MyData, is that it starts with the idea that individuals can manage and control personal data because they have ownership.

Web 3.0 has a similar view on this. Web 3.0 is not yet

clear. Still, it is also defined as a "personalized intelligent web ecosystem" that can read, write, own, open, and distribute through blockchain-based identity authentication. Simply put, it means a world where individuals can manage and distribute digital content or assets owned by themselves without going through big tech platforms. Here, we find something in common between MyData and Web 3.0, which is, after all, self-determination. MyData talks about the right to self-determination of personal data and the right to self-determination of digital assets. Personal data is also part of digital assets if you look at it broadly. In the end, MyData and Web 3.0 will both move within the digital economy, and they are pursuing an ideal world in which individuals' right to self-determination is guaranteed, whether it is an Internet- or a blockchain-based world. So, what will the world look like when MyData and Web 3.0 are combined in the future? In short, likely, the real economy's MyData and the virtual economy's Web 3.0 coexist, and the world is clearly expected to guarantee the right to self-determination of personal data more than it is now.

Web 3.0 is often mentioned when talking about data sovereignty. Tim Berners-Lee, the founder of the Web, pointed out the current problems with the Web, saying, "We demonstrated that the Web had failed instead of served humanity, as it was supposed to have done, and failed in many places" In other words, the web originally started from guaranteeing the rights of web participants along with efficient sharing of information. However, entering Web 2.0, this original purpose was lost, and it can be said that the flow to solve it again began in Web 3.0. Currently, Web 3.0 is a new type of web behavior model in which service participants share profits, and it is developing into a decentralized web in which individual users own and protect their data and personal data without platform dependency. This means that personal data, including personal data, is not stored on a centralized server provided by a specific platform operator but is distributed and stored in online data stores or cloud services selected by individuals. The core philosophy of Web 3.0 focuses on openness, spontaneity, and transparency. All data, including web users' personal data, presents principles such as guaranteeing data sovereignty wholly owned by the user and free participation without major regulatory agencies. In summary, Web 3.0 can be defined as a web in which web users' data, personal data, etc., are not subordinated to the platform but in the form of sovereignty over data given to users.

Blockchain & DLT (Distributed Ledger Technology)

Blockchain is considered the next-generation revolution in financial services around the world. Additionally, blockchain technology is expected to be used in some form in the MyData industry because of the decentralized characteristic of technology, which matches well with the human-centric idea of MyData. I want to point out the basics to help you understand blockchain technology's components and how they work. The technology is expected to be used in many ways in everyday life, including creating smart contracts, improving payments, maintaining medical records, and digitizing identity. Cryptocurrency is one of the applications of blockchain. This new form of currency can make payments cheaper and eliminate the need for money (physical or paper). In recent decades, we have seen how banks can be mismanaged. Trust in investment groups has proven to be a structure prone to exploitation. Ultimately, people were looking for a money trading system that prevented fraud. Distributed ledger is an innovative technology that helps solve the current problems of many financial systems. Trust is essential for financial transactions, and one of the reasons why banks are causing problems is a lack of trust in each other. Through a distributed ledger, people can trust each other completely.

One of the most remarkable features of blockchain is smart contracts. General paper contracts are characterized by legal expression and rely on a third party, the public justice system, for the execution and dispute situation of the contract. On the other hand, intelligent contracts execute contracts using programmed computer code. Smart contracts can eliminate the need for automated and self-executable intermediaries. By programming certain conditions, the contract can be performed independently, for example, by imposing a fine in the event of a specific event. This is the most significant area of change. Using blockchain as a tool for contracts will cause many changes. Blockchain allows you to record something in a shared ledger. Once registered, the transaction is displayed in the database, providing irrefutable digital evidence that the transaction occurred on a specific date between the two parties and completing the implementation of the contract in the digital world. This area requires a lot of research in the field of contract law. This is because there is room for interpretation to change by comparing smart contracts with actions such as subscription, consent, establishment, implementation, and termination of existing legal contracts. One of the promising applications of smart contracts is the music industry. If you set up a blockchain at the time of the release of the sound source, you can sell the good source and distribute benefits in real time between multiple parties in the value chain. Many areas, such as securities, syndicated loans, trade finance, swaps, and derivatives, are also available.

A distributed ledger is a type of database. A key feature is that it has data replicated and shared across multiple sites, countries, or institutions. Since this data is consensually stored, there is no doubt about the reliability of the data, and it is generally disclosed. Instead of grouping transactions into blocks, records are stored continuously in a continuous ledger, and new blocks can only be added after the participant has reached a quorum. Using a distributed ledger, you can safely share data in real-time (there can be multiple differences here). There are many types of distributed ledgers; the main difference is how consensus is achieved. Examples of distributed ledgers include Ripple, Multichain, and Hyperledger Project. The ledger may be open or private. The Public Ledger can provide data to anyone, and all participants can see a copy of the same ledger. An example of this is Bitcoin, a database against censorship. Private Ledger can distribute the exact copy as the ledger, but only to a limited number of participants.

Blockchain is a data structure that can provide agreement and security when sharing data. It may be the safest database model for building financial transactions in the digital world. A common concept for existing databases is that they are stored on a single server with a delegated server to manage the database. In contrast, blockchain databases have distributed systems with many interdependent computers involved in database management. Therefore, it is virtually impossible to hack a database because a single computer is unreliable.

In principle, a distributed ledger is more difficult to attack because multiple shared copies of the same database must be attacked simultaneously for a cyberattack to succeed. However, suppose someone can find a way to "legally" modify one copy. In that case, it can change all multiple copies of the ledger, so this does not mean that distributed ledger technology is not vulnerable to cyber-attacks. Blockchain systems generally consist of two main components. These two components are a P2P network and a database. Concerning networks, a blockchain consists of a group of computers connected through a communication model called a P2P network. This is a mechanism by which computers communicate new changes to that database. The second major component of the blockchain system is the database itself. A database is an accumulation of transaction history. The system allows transactions to be recorded in the order in which they occur. Looking more closely at these two components, a P2P network comprises many computers called "nodes." These node networks are simply connected to each other. There is no limit to the number of nodes connected within the network. Today, there are networks with thousands of nodes connected to the chain, and the total number of nodes increases or decreases daily.

Connecting these nodes means that there is no single point of failure because a single computer cannot cause transaction failures. Because all other nodes can simply propagate the information in the database, a single node in the network cannot hack the database or sensor information in the database. When a new transaction appears, it can be sent to any node, and it doesn't matter which specific node it is. Transactions are typically transferred to many nodes from the start and are challenging to track. When a node receives a transaction, it sends it to all adjacent nodes, etc. Transactions propagate quickly across the entire network, which is how blockchain works in a P2P network. These nodes are in various locations, and transactions appear in multiple locations at once, so the source of the transaction is not known exactly. It is difficult to censor transactions in this way. When a transaction is accumulated, it collects the transactions and begins.

There have been many projects in the blockchain community, and the most notable project was the DAO (Decentralized Autonomous Organization) in Ethereum. DAO was an open-source, leaderless organization with contracts stored in the Ethereum database. By 2016, DAO had collected contracts worth $150 million from Ether. But it was attacked in June and lost 3.6 million Ethers. With this attack, the Ethereum community voted. Most participants agreed to change the Ethereum code to recover the stolen funds. However, a small number of users disagreed. Minorities argued that there should be no unauthorized change in the blockchain to make a sound history. So, they kept mining the previous version of the blockchain. This is how the platform is divided into Ethereum (ETH) and Ethereum Classic (ETC). The Ethereum Classic is originally a continuation of the Ethereum blockchain. The history remains unchanged, and there is no external interference. In contrast, Ethereum (ETH) is a blockchain that has been hard forked by taking steps to restore funds to its owners. It is worth noting what can happen when members of this type of decentralized organization make different value judgments in the personal data ecosystem.

Blockchain is expected to contribute to the decentralization and further democratization of the global financial system. Since everyone is given equal access, financial institutions should recognize the importance of blockchain in the financial industry. As technology advances, financial authorities are paying attention to the enormous impact of blockchain in all aspects of their daily lives. Blockchain can eventually render the traditional methodology of conducting transactions obsolete. Whether trading assets, values, stocks, options, or derivatives, this technology can create a more efficient and flexible financial system. For example, many redundancies and mistakes are eliminated from post-transaction agreements, saving vast amounts of money paid to trusted third parties and helping to make the global financial system more efficient. Of course, this is a story in a limited area. Still, it seems clear that blockchain in finance will expand with changes in technology and institutions, and the next area will be the personal data industry.

PDS (Personal Data Stores)

Personal Data Stores/Services (PDS) can be defined in many ways, but Wikipedia's definition is the most value-neutral form of definition. A PDS can be defined as a service or repository that enables an individual to store, manage, and distribute personal data in a secure and structured manner. Defining a PDS only briefly makes it challenging to convey its implications sufficiently, so the purpose or characteristics of the PDS are usually also mentioned. It is commonly discussed now that users should be given "control" over their personal data. As the PDS itself is perceived as a 'storage,' it always prioritizes the storage of data but often overlooks that 'storage' itself is not the purpose of the PDS. Ultimately, the realization of self-determination over personal data is key, and data storage is not important. The data storage location for PDS is irrelevant, such as a location, an external distributed storage, or a combination of the two, and the user's terminal can also be a storage.

Perhaps it can be said that PDS itself started earlier than the history of MyData. There were various PDS projects such as Data, The Locker Project, open PDS/SafeAnswers, Higgins, and ID Hole. These early PDS projects suddenly faltered as ID Federation services became more common. This is because the ID Federation service, commonly known as the social login service, served as a PDS providing personal data. We provided sufficient personal data (such as user identification and distinction, email address, etc.) at that time. Of course, many services these days do not provide enough basic information. For example, in the case of Airbnb, additional information (address, payment information, etc.) is needed even if you log in to social, so it is not enough as a basic ID Federation service, so PDS or similar services are increasingly likely to be used. In the past, inactive PDS projects and OpenPDS/SA (SafeAnswer) projects were similar to the concept of decentralized identification in that they did not deliver the data stored in the PDS as it was; only the answers were needed.

The concept emerged in the early 2010s, and many of the current inactive projects are Silo Base PDSs. At that time, it was not well received. It artificially created another big brother when there was already concern about personal data held by global conglomerates such as Google. It's mostly a fade-out architecture with the advent of the ID federation service. It's the same PDS in the server and client architecture, but cloud-based PDS has been deployed slightly differently. The reason is that in the case of the cloud, the PDS service provider is interpreted as a processor rather than a controller based on the EU GDPR. Especially if you're using storage space in your private cloud, it emphasizes all aspects of allowing individuals to control all of their data because it is easy for people to recognize the concept of collecting and storing all personal data in one place. These include Digi.me, Inrupt (Solid project), Mydex, and Japan's Information Bank (DPRIME).

PDS, an infrastructure for managing personal data, can be operated in many forms. Initially, developments were made as a separate connection between the information provider and the recipient (API model). Then came a platform for collecting and distributing data from multiple data sources (Platform model). Both models are essentially corporate and institutional-centric, built and operated primarily on the organization's needs. However, the basic idea of MyData is the point of integration between organizations that hold and use personal data, with people at the center of the ecosystem (Operator model). In the Operator model, personal data management service providers form an interoperable ecosystem and compete for infrastructure for personal data transfer.

Because MyData is oriented toward individual-centered services, in MyData, individuals are not limited to the general concept of users of existing ICT services. Conceptually, individuals are closer to service providers' role in data flow because they act as content providers that generate personal data. PDS architectures that faithfully reflect this concept have begun to emerge, and packages that allow individuals to utilize their smartphones or even operate servers that individuals themselves host to store their personal data have emerged. For example, there is a Debian-based package distributed by CozyCloud to serve as server-based self-hosting, and there is SNPLab Inc., which serves smartphone-based on-device PDS models, which will be described later.

Authentication

Authenticating identity through personal identification is often the start of the MyData service. A third party usually does identification. However, decentralized identity verification technology can be utilized for identity verification if user authentication is not the main purpose but identification in the verification area of information is.

The user authentication procedure is the process of identifying whether you are a legitimate user to allow access to the data, and the general authentication techniques can be classified into four main categories, depending on the type of user authentication. The first is Knowledge-Based Authentication (KBA), which includes ID/PW, PIN, and question-and-answer authentication. The second is using the authentication means in possession, such as OTP, mobile phone SMS, and public certificates. Third, methods of authentication using biometric information such as fingerprints, iris, veins, etc. The fourth category is authenticating by analyzing user behavior patterns such as signatures, keyboard typing patterns, and gait; they are classified as behavior-based authentication. It is necessary to determine which user authentication to use and how to use it by selecting the appropriate method for implementing the service, and each user authentication is as follows.

Knowledge-based authentication is a user authentication method based on secret information previously set and shared by a user and a server. It is characterized by using the knowledge that the user remembers. Identifying commonly used IDs and passwords to authenticate users is a typical example of knowledge-based authentication (passwords, PINs, etc.). Knowledge-based authentication does not require additional hardware, easy system deployment, and user convenience. However, the disadvantage is that authentication is less intense than other methods, making it vulnerable to security. Attackers can often infer it, and in terms of usability, it is easy for users to forget, and there is a high risk of leakage, making it vulnerable to social engineering.

Ownership-based authentication is a method of performing authentication using user-owned verification means (co-certificate, token, smart card, OTP, etc.). While hardware-type tokens are less portable and convenient because users must own physical-type tokens, software-type tokens can compensate for some disadvantages of hardware-type portability and convenience. This authentication method is more secure than knowledge-based authentication because the user must have authentication means by default. However, there is an inconvenience that users always have to own for authentication, and there is a high risk of leakage when stored on a storage medium in the form of software and files. Tokens stored in software or files are not used alone because of the risk of replication, loss, and theft and are usually used in conjunction with other authentication methods.

Bio-based authentication and behavioral-based authentication are authentication methods that utilize the user's unique biological characteristics and can be divided into a method that utilizes physical and biological characteristics and a method that utilizes behavioral characteristics. The physical authentication method extracts different biometric information (fingerprints, iris, face, vein, etc.), registers the information in the system, and determines certainty by comparing the biometric characteristics measured through the biometric input device with the registered information. Behavior-based authentication methods can utilize voice recognition, signature pattern recognition, and gait by authenticating the user based on the behavioral results that appear when the user uses the body. The advantage is that the user does not have to own a separate authentication token, and there is no information to know separately, so it is highly convenient. Moreover, it is a user authentication method that has been widely used recently because it uses unique information from the user's body, which is highly secure and difficult to replicate. However, since it is a system that utilizes biometric characteristics and pattern information, the cost of construction is high, and biometric information can change depending on an individual's health status and verification environment, so there is always a False Acceptance Rate (FAR) and False Rejection Rate (FRR). As a result, it is difficult to use it alone because others may mistake it for me or reject it, so it is usually used with knowledge-based or ownership-based authentication.

These user certifications are divided into four levels of Level of Assurance (LoA) by ISO/IEC 29115 to separate authentication requirements and roles, including authentication threats and threat controls, warranty level criteria, associated entities, and user authentication processes. Assuming MyData will evolve to a combined type, various authentication measures should be considered simultaneously.

SSI & DID

Some of the keywords that always appear when referring to MyData are SSI (Self-Sovereign Identity) and DID (Decentralized Identity). That's because the identity authentication method is closest to the ideal MyData pursues. SSI, mainly translated as "self-sovereign identity," is a form where individuals directly manage personal data independently, unlike existing user authentication by service providers or third-party certification agencies. In the early days, most user authentication occurred in a centralized environment for each service. This includes the Silo model mentioned earlier in the PDS. Then, with the emergence of IDP (Identity Service Providers) such as Google, Facebook, Naver, and Kakao, Federated Identity, which is serviced in the form of social login, is widely used. These identity authentication services inevitably concentrate personal data on companies, institutions, and IDPs, so there is a problem of relying entirely on the companies, institutions, and IDPs for the ideals MyData pursues.

This identity authentication environment in which individuals directly manage and utilize sovereignty requires the implementation of an identifier that can distinguish individuals and an ecosystem that utilizes identity data corresponding to the attribute data of this identifier. Blockchain-based decentralized identity verification (DID) is mentioned as an identifier generation and management technology for a self-sovereign identity environment. DID is widely used as electronic identification, allowing users to store personal data on their terminals (e.g., smartphones) or specific spaces and select and submit only the necessary information. For example, when purchasing alcoholic beverages at a mart, you usually show your resident registration card. Still, all personal data such as address, name, and resident number is exposed in this case. But what's needed is to check whether you're an adult. DID-based identification wallets are often cited as personal data utilization mechanisms that achieve the desired purpose while preventing excessive personal data leakage by checking only necessary facts like "adults aged 20 or older" and are meaningful as SSI implementations in terms of returning data sovereignty from institutions and companies.

Scraping

In short, scraping is a data extraction technology. Specifically, it refers to a technique automatically extracting data from a web page or other program screen. Much scraping on a website means extracting the data you want from a website or web page. This process uses automated data structures using computing devices to transfer data between programs. Often, scraping objects are constructed in difficult forms for humans to read, distinguishing them from simple parsing. This is because it is often assumed that the output being scratched is displayed to the end user. Scraping often includes information that hinders automated processing, such as binary data, images and multimedia data, specific formats, labels, and annotations. It is primarily used for downloading and collecting MyData in many forms of transmission.

Scraping can be largely divided into web scraping and screen scraping. Web scraping is a method of running a browser to collect on-screen data. Sends an HTTP GET request to download content to a specific website before extracting the desired information. This is usually called a scraper bot, and when the web server responds to this HTTP GET request, the scraping algorithm analyzes HTML documents to extract data with specific patterns and stores the extracted data in a database for use in any form. Web scraping is used in various areas where certain automatically-collected information is needed. In the financial and stock markets, scraping technology is used to gather news information, and analysts automatically collect information on corporate financial statements for investment advice. In the e-commerce market, scraping technology is also used to collect information on competitors' products and quickly identify price fluctuations. Financial institutions also use it to collect data from the government's website and the National Tax Service's website to identify credit information before lending. In the sense of collecting the desired data, the term crawling is also used in a similar sense, which is often used technically. However, the difference is that crawling continues to crawl around like the word's meaning, but scraping only targets certain pages.

Originally, screen scraping was a technique for reading text data from the screen of a computer display terminal, which meant programming the collection of visual data from sources instead of parsing data, usually like web scraping. This was typically done by reading the terminal or the terminal's memory. At the time, it was implemented by using an auxiliary port or by connecting the terminal of the computer system or the output port of the terminal to the input port of another computer. Screen scraping often means a two-way data exchange, as control commands monitor the user interface or control programs that enter data into an interface intended for human use. The financial sector performs positive input/output and data exchange in screen scraping scenarios. Sometimes, screen scraping includes web scraping, but even when scraping a web screen, it usually means scraping directly through the HTTP communication module without running a browser. It is sometimes classified as protocol-based scraping to distinguish it from web scraping. Protocol-based screen scraping is mostly used because it has advantages in terms of speed, independence, and convenience to process it in a protocol manner without running a browser.

Scraping technology can extract data from various sources, including the web. It's different from crawling because deduplication is not always necessary. After all, it's a method of extracting specific data. Crawling is a method of downloading pages and links from the Web. Deduplication is essential because it only works on the web and does not recognize that the same content has been uploaded to multiple pages. With these advantages, screen scraping has become one of the most used technologies in the fintech sector. In addition, the protocol method has the advantage of developing and maintaining scraping separately from the user interface. Still, there is a high possibility of service failure when changing the site. Simply put, a separate response is often required when a web page is reorganized or changed. Despite various disadvantages, scraping technology is widely used for MyData because it is a technology that can easily download one's personal data directly according to the individual's will.

API

API is an abbreviation for Application Programming Interface, which means a set of definitions and protocols for building and integrating application software. It refers to a connection between a computer or computer program, which is itself a kind of software interface and serves other software. APIs for MyData services, for example, financial MyData, can be broadly classified into authentication, information provision, and support functions. The authentication API is an API that users need to request the transmission of personal credit information and perform self-authentication. It is once again divided into individual authentication and integrated authentication. Individual authentication is a self-authentication method that uses an information provider's autonomous authentication method to authenticate as many information providers as the number of information providers that a customer requests to transmit personal credit information. The information provision API is an API necessary for information providers to transmit personal credit information to financial MyData operators based on a user's request to transmit personal credit information (bank, insurance, card, financial investment, electronic finance, etc.). In addition, the APIs required to support the services of financial MyData providers can be classified as support APIs.

Basically, API standardizes all access, so anyone promises the same access as long as the conditions are right, regardless of device/operating system, etc. In addition, functional APIs can be used when developing applications in an organization to easily use the basic functions required–authentication, communication, payment processing, and location verification–without having to self-develop/update every time. It has the advantage of simplifying development workflows and easy application expansion. It is easier when you want to build a flexible service environment and integrate software and when you need collaboration among developers. Although API methods are usually described as superior in terms of security, it is difficult to say superiority and inferiority because API gateways, which are a single entry point for APIs, can be targeted by hackers. The advantage of accessing using ordinary HTTP methods may be a great disadvantage when it comes to security. However, standardization and development costs are more burdensome than that. One of the biggest drawbacks to the design of REST APIs is that in the absence of this formulated standard; it is difficult to manage and actually expensive to implement and deliver API functions in terms of development time, ongoing maintenance requirements, and support delivery. Though API itself is not the technology only for MyData, it is often mentioned in terms of implementation.

De-identification

There have been large and small leaks in utilizing personal data, and social demands have continued to strengthen privacy policies. Therefore, when personal data is collected and used in one place, whether a company or an institution, compliance with related laws for use and the act of collecting them in one place should be preceded. In practice, separate consent for collection and utilization is essential. Of course, data utilization for statistical and research purposes other than marketing and other business purposes is possible without consent. However, this also has the premise that it should be used in a form that can identify individuals, that is, in a form that is not personal data. This requires de-identification technology. De-identification measures include deletion, pseudonymization, total processing, data deletion, data categorization, and data masking, which can be utilized alone or in combination.

First, 'deletion' is the most basic and powerful de-identification process, mainly for identifiers. We create de-identified data in non-personal data form by deleting values or names uniquely assigned to individuals or objects related to them. Of course, data reduction is a principle if the attribute value, not the identifier, is not related to the purpose of data use. The attribute can also act as a quasi-identifier in certain environments. There are various de-identification technologies other than deletion, and representative techniques are as follows.

Pseudonymization is the first technique mentioned during de-identification processing that replaces personally identifiable data with other values that cannot be directly identified. The low data deformation or deterioration level is due to replacing names and other unique features (school of origin, workplace, etc.) with other values. Still, the disadvantage is that the unique attributes that can be identified continue to be maintained when given alternative values. Among them, the Heuristic Pseudonymization technique can replace the values corresponding to the identifier with several set rules or be processed according to human judgment to hide detailed personal data. Because all data is processed in the same way without considering the distribution of identifiers or prior analysis of collected data, there are limitations in the alternative variables that the user can exploit. Some vulnerabilities expose you to certain rules that are replaced with other values. Therefore, careful consideration is required when establishing rules to ensure that individuals are not easily identified.

Encryption is a method of replacing personal data by applying algorithms of certain rules when processing data. Usually, we have a decryption key so that it can be decrypted again, so we need a security measure for it. However, decryption is fundamentally impossible in the case of one-way encryption such as hash. One-way encryption completely removes the identification of personal data, which is a more secure and effective de-identification technology than two-way encryption, but it is still considered identifiable.

Swapping refers to exchanging existing database records with predetermined external variable values, and aggregation is applying statistical values (all or partially) to prevent identifying a particular individual. Sensitive numerical information can be unidentified by applying date information related to an individual (birthday, qualification date) and other unique characteristics (sensitive information such as physical information, medical records, medical history, and specific consumption records). Although it has the advantage of creating data sets for statistical analysis, it also has the disadvantage of being able to identify by inference if it is difficult to analyze and the aggregate quantity is small. Total processing is divided into processing the entire data and only the partial total. Re-alignment, which applies up, down, and rounding or prevents specific information from being linked to the individual without compromising the entire information, is also a de-identification technology.

Data Suppression is a technique that prevents individual identification by converting and/or categorizing specific information into representative values for that group or converting and/or categorizing into interval values. It has the advantage of being able to analyze and process information that can identify individuals (address, date of birth, unique identification information (resident registration number, driver's license number, etc.), and user accounts (registration number, account number) using statistical data format. Data Masking refers to a technique that converts all or part of the information (name, phone number, address, date of birth, photo, unique identification (resident registration number, driver's license number, etc.) and user accounts (registration number, account number and email address) into alternative values (blank or noise). It is possible to remove the personal identifier. It has the advantage of fewer variations on the original data structure. Still, it is difficult to utilize for data-required purposes if masking is applied excessively, and it is possible to infer specific values if the masking level is low.

After de-identification processing, it is necessary to evaluate whether the de-identification process is appropriate, and this method of evaluating adequacy is called the privacy protection model. This is because there is a risk of individuals being identified through combination with other information, such as public information and various reasoning techniques if there are not enough de-identification measures. Typical types of models are as follows. First, k-anonymity is a privacy protection model proposed to defend against vulnerabilities such as linkage attacks on published data. It allows at least k equal values to exist in a given data set, making combining them into different information difficult. Second, ℓ-diversity is a model to defend against two attacks on k-anonymity, homogeneity attacks, and background attacks, and records that are de-identified together in a given dataset have at least ℓ different sensitive information within the homogeneous set. The third t-closeness is a privacy model to compensate for the vulnerabilities of the ℓ-diversity, which considers the meaning of values. Assuming that any 'same set' consists of 99 identical records and one different record, an attacker can see that the target is the same record with a 99% chance. Similarity Attack's vulnerability is that privacy can be exposed even if it is de-identified through the ℓ-diversity model, assuming that the information in the de-identified records is similar. Thus, it is a privacy model in which the distribution of specific information in a homogeneous set and the distribution of information in the entire dataset show a difference of t or less.

Conceptually, MyData's main service is multiple information providers and scenarios that combine data from different industries. Therefore, it may be difficult to regard one technique as sufficient de-identification measures, so it is necessary to select and utilize appropriate techniques and detailed technologies considering the purpose of using personal data and the advantages and disadvantages of each technique.

Other Notable Privacy Technologies

In addition to the de-identification technologies, the Future of Privacy Forum has presented ten privacy-enhancing technologies companies should pay attention to over the next ten years. These are ZKP (Zero Knowledge Proof), Homomorphic Encryption, SMPC (Secure Multi-Party Computing), Differential Privacy, Edge Computing, Device-Level Machine Learning, Identity Management, Small Data, Synthetic Data Sets, and Generative Adversarial Networks. Those technologies are desirable to apply when implementing MyData services, and the most important and promising technologies for MyData are as follows.

Homomorphic encryption is a technology that allows data to be calculated while encrypted. In 1978, the first concept was presented by Rivest, Adleman, and Dertouzos, and it was a cryptographic system that could perform both addition and multiplication while encrypted. So, when describing homomorphic encryption, we often compare it to processing gold using a glovebox. When the owner of the gold wants to leave it to the artisan and process it into a gold ring, the artisan may steal some of the gold or replace it with something else, so let the craftsman proceed with the gold in the glove box and lock it. It is similar to a series of operations in which the owner opens the lock and pulls out the processed gold ring after the operation. It encrypts data with isomorphic cryptography, which is processed and analyzed by the side that processes it. Still, it cannot be extracted or replaced or retrieve anything about the data. However, the biggest issue with isomorphic passwords is processing speed. Several attempts are being studied for practical commercialization because it is processed at a much slower computational speed than processing data without encryption.

Differential privacy is a privacy model that prevents an attacker from obtaining more information about a particular individual, whether or not a particular individual is in that database, through the response values of the query. The de-identification models have the advantage of being relatively easy to apply and verify. Still, they pose a problem: the risk varies depending on the type or form of an attacker's background knowledge. No matter how robust de-identification algorithms are applied, an attacker will always know sensitive information in the worst case. In other words, the k, ℓ, and t privacy models alone are not enough to be a universal standard for personal data protection. Differential privacy is proposed as a privacy model that overcomes the limitations of de-identification techniques. The data disclosure mechanism makes it difficult to differentiate between data sets with and without arbitrary records, making it difficult for an attacker to acquire sensitive and meaningful information about any data in the worst-case scenario. Accordingly, differential privacy is considered suitable for application as a universal privacy criterion.

Zero knowledge-proof technology needs lots of mathematics but is not difficult to understand conceptually. When someone proves to the other person that a statement is true, let's assume that it is guaranteed nothing will be exposed except whether it is true or not. The person who wants to prove that a sentence is true is called a Prover, and the person who participates in the proof process and exchanges information with the proof is called a Verifier. Usually, we use circular cave models to explain easily. Assuming that there is a circular cave with one entrance and a door blocking the cave on the other side of the entrance when the attestor has the key to a door in the middle of a cave and must prove to the verifier that he has the key but tries to prove it without showing it to others, the proof of key retention is possible in the following procedure. First, suppose that the attestant enters the cave through random passages A(upper) and B(lower), and the verifier cannot see which passage the attestant entered outside the entrance. If you ask the verifier to come to either A or B, if the verifier has a key, the verifier can come to the path no matter which path the verifier chooses. However, if the attestor does not have a key, it can only come out through the passage where it first entered, so it cannot come out as requested by the verifier with a 50% probability. If you repeat this multiple times, you will have a high probability of determining whether you have a key. Proof of thumb is that you can prove specific content in this way without exposing personal data. Proof of identity is mainly used for anonymous coins that value personal data protection, and information other than that disclosed by the transaction provider in cryptocurrency transactions is used to make the recipient unknown.

Federated Learning can be considered a type of edge

computing and device-level machine learning, and there can be many use cases based on the privacy-enhancing technologies mentioned above. The higher the convenience, the higher the likelihood of security and privacy issues. Therefore, when providing MyData services, it is necessary to utilize these various technologies that can protect privacy.

VIII.
Suggestions for Personal Data Sovereignty and Utilization

Background

The starting point of all problems related to personal data begins with the collection, use, and management of personal data owned by individuals by unauthorized companies and institutions. Therefore, we arrive at the simple but difficult conclusion that the most fundamental approach to solving personal data-related compliance problems is not to provide personal data. Blockchain has become a hot issue in the IT industry because it is a thought change. It began with a shift in preventing forgery by widely distributed storage rather than hiding important ledgers deeper. As a result, scenarios have been created that can solve difficult problems with existing mechanisms at once. Of course, blockchain cannot solve all problems. There are limitations, but there are clear advantages in certain scenarios. In addition, limitations have gained a lot of attention as a technology and idea that will change the future because at least some of them are often overcome with the development of technology.

The basic structure of the MyData platform proposed in this book is based on storing personal data only on an individual's device, such as a smartphone. It may not be new to store personal data only by individuals, but adding a few things can be a new idea. The technologies in Satoshi Nakamoto's Bitcoin paper were not new either. Hashing techniques and POW (Proof of Work) algorithms were all existing technologies. Likewise, although blockchain is no longer a new technology, it can create a completely new idea by adding the idea of allowing only individuals to store personal data. This chapter will present new suggestions for the personal data use paradigm and logically demonstrate the advantages of implementing this change of idea in a data use ecosystem. In particular, when dealing with combined finance and ICT data, such as fintech, information security accidents, and personal data protection problems cannot be avoided. Despite the risk of leakage from personal data collection, many companies, including fintech startups, still want to collect personal data because of the value generated from personal data. Companies try to collect personal data despite information leakage and compliance risks because statistical analysis can find profitable markets and generate more profits from target marketing.

On the other hand, individuals do not particularly benefit from providing personal data to companies. Rather, personal data is leaked, resulting in the spam or even exposure to various phishing risks. This unfair and one-sided relationship is neither healthy nor sustainable. The proposal in this book is more meaningful because it is not a one-sided relationship but a more fundamental change, that is, a paradigm shift, rather than a complex solution for individuals and companies to become a win-win healthy ecosystem.

Utilization of Personal Devices

In the early days of the Internet, individuals did not have a device to manage personal data, so the service began by providing personal data to create an account, but at this time, mobile devices are common; there is a scenario in which services can be provided without providing personal data. An example is using the ID Federation service for user identification without a separate subscription procedure. Numerous services have long been able to provide services immediately without the user subscription process and collection of personal data by signing in with Google, Facebook, etc. You only need to provide user identification using your Google account so that game characters can be distinguished. Companies don't need to collect other personal data (of course, for games with age restrictions, additional information on whether or not age conditions are met) to provide game service. In general, individuals provide personal data to companies when they join the service, but over time, they often do not remember what data they gave to which companies. If the timing of the subscription is different, the personal data kept by the company will be different. There is one individual and one address, but commonly, the address differs depending on the subscription time. What's more inconvenient is that for sites that have been subscribed for a long time, changing addresses requires a tremendous effort to find an ID or password you don't remember. We will focus on fundamentally solving these points and leave the implementation problem as a future task.

Over the past 20 years, IT has made remarkable progress, and now artificial intelligence has begun to replace human work. Still, personal data is managed in the same form as 20 years ago. This was inevitable in the early days of the Internet when individuals did not have devices or equipment to manage personal data. Still, now that most people have mobile terminals (e.g., smartphones), they can store personal data in their IT devices with storages. In particular, unlike general PCs, smartphones are the most personalized devices and are more suitable for storing personal data than any other device. Although this article is limited to smartphones, storing it in a private cloud can have the same effect, and in some cases, backing it up to a private cloud can be more advantageous in terms of personal data management. However, the mechanism proposed in this book will be storing personal data in mobile terminals, the most personalized devices, and allowing companies to use personal data only within individual mobile terminals.

Implications of Personal Data Stored by Individuals

Recognizing that an individual owns personal data has important social, legal, economic, and technical implications. First, the data held directly by individuals is technically superior. As mentioned earlier, user-held data represents the most accurate data about an individual. It's the latest data. No organization can have data as accurate and reliable as itself. Second, if the user holds the data, it is legally superior. In the user-centric model, an individual grants access to data to a third-party service provider through prior explicit consent. The perception of individual ownership of data solves many of the problems that companies face, and there are few legal restrictions if they utilize personal data without storing it. As a result, organizations comply with data minimization requirements and can legally access and process all data. Third, user-held data creates opportunities for companies to build new types of applications based on user-held data. This enhances the security and privacy of digital interactions. The possibility of data infringement is reduced because customer data stored on centralized servers is reduced. In addition, user-owned data allows companies to make more personalized proposals and provide customized services to customers. Finally, the form in which an individual directly owns data can have a socially desirable impact. If practical tools are developed that allow individuals to control personal data in detail, individual data utilization will increase. In addition, interactions with service providers will be more equal because individuals will set conditions for third parties to access the data. This is likely to contribute to a change in the perception of the value of personal data. Personal data will emerge as assets that create long-term value, not disposable.

Furthermore, if we use our data so that only individuals store personal data and companies and institutions do not, a new type of data use ecosystem is possible, completely different from before. Similar attempts have been made through open platforms such as open PDS/SA (Safe Answer). This attempt to reverse the idea of changing the basics can bring about new changes and, in some cases, become a groundbreaking fundamental solution. In countries that selectively refuse (opt-out) organ donation, very few people die waiting for an organ donation. We need to pay attention to the case that by changing the ecosystem's default to refuse organ donation rather than to apply for organ donation, organ trafficking, and human trafficking have decreased beyond the targeted social problem, the transplant organ shortage.

Until now, personal data has always been given to companies by individuals. We remember that the first thing to do to get a service was to generate an ID. Generating an ID itself is already providing an identifier for me. If even such an identifier is a platform only I have, almost all kinds of personal data-related compliance will be resolved. Privacy-related acts are based on the fact that personal data does not belong to companies or organizations. If all personal data, including such identifiers, are kept to individuals, almost all kinds of existing personal data-related compliance will be resolved. Privacy-related regulations presuppose that personal data does not belong to companies and organizations. If individuals store and process personal data directly, such problems will be fundamentally solved.

On-Device PDS

In previous chapters, on-device PDS was introduced as a special form of PDS, along with many other kinds. On-device PDS are separated in this book because a completely different approach to implementing MyData is possible based on a fundamentally different form from legacy PDS. Another reason is that despite the difficulties of technical implementation, on-device PDS is considered the best MyData platform to guarantee personal data sovereignty.

The term on-device comes from on-device AI. With the advent of the Fourth Industrial Revolution, there is a high interest in artificial intelligence (AI), which learns and deduces like humans. In particular, AI technology will be widely applied to the Internet of Things in smartphones, robot vacuum cleaners, and washing machines in the future. On-device AI, attracting attention as a next-generation deep learning technology, is artificial intelligence experienced in my hands. AI technology has been carried out by transmitting information collected from personal devices such as mobile phones to a central cloud server, analyzing it, and sending it back to the device. On-device AI, on the other hand, can collect and calculate information on smart devices themselves without going through remote cloud servers, as its name suggests. As such, on-device AI processes information inside the terminal device, enabling fast operation with low latency. Because it is not through a central server, it can also solve the security problem that has emerged as a problem with cloud-based AI. In addition, on-device AI, which is associated with edge devices, which are hardware that comes in direct contact with users, has the advantage of making decisions that are more appropriate for the actual user because it directly experiences the environment and processes information.

Unlike in the past, PDS is required to go beyond simply storing and managing personal data, and it is expected to eventually develop into the same form as an artificial intelligence secretary based on personal data like J.A.R.V.I.S. of Iron Man. In the case of Apple, instead of AI services that connect LLM (Large Language Models) to the cloud, such as Microsoft and Google, it adopts an on-device model that runs separately on products. In WWDC 2023, Apple mentioned AI features in its products instead of talking about specific AI models, learning data, or how to improve in the future. The strategy is to permeate AI throughout the product without putting AI technology at the forefront. Since it is a company based on iPhone hardware and its ecosystem, it prioritizes iOS-based device sales rather than reorganizing search or improving productivity software like Google. Once these AI models can be run on mobile phones, they need less data, and privacy issues can be avoided. This applies equally to PDS scenarios in which sensitive information is combined and stored.

Suggestions for the Use of Personal Data by Companies

Companies and organizations often use personal data rather than individuals. In this session, a new personal data utilization ecosystem is proposed to solve problems in existing ecosystems, such as the administrative burden of collecting, managing, storing, and destroying them according to legal requirements with consent. When personal data is transmitted to the data itself, replication is free due to the nature of digitized data, and there is no way to limit its use. Therefore, design in delivering the data should be avoided as much as possible when using personal data. In this suggestion, the right to use personal data, not personal data, is digitally assetized and stored in the blockchain as tickets and/or tokens. The proposal that specifies the type, period, number of times, compensation, and purpose (statistics and marketing, etc.) and the permitted entity/service will be posted on blockchain by the company/organization, and the individual will approve the proposal. When approval is confirmed, the proposal becomes a right to access personal data, and the company/organization uses the personal data in the device as requested. It is a concept similar to the permission mechanism in smartphone OS and can be understood as an extension of general data beyond the device. It can be seen as the provision of access rights that are digital assetized in terms of period, number of times, compensation, and purpose for the personal data.

The main stakeholders operating on this platform are companies/organizations, individuals, and foundations to operate the platform. Companies and organizations are mainly subjects that request personal data, and they post proposals that specify the period, number of times, and purpose of use. Individuals directly manage personal data, respond to personal data usage requests through a blockchain-distributed app (De-centralized App), and receive compensation for using their personal data. In contrast, the personal data is processed and controlled by the individual. The foundation can manage the Root Certification Authority (Root Certification Authority) for certificate issuance and issues certificates to participants by utilizing the Web Portal for the platform, HSM (Hardware Security Module), and DLT (Distributed Ledger Technology). To maintain the data economy ecosystem, a reward pool can be operated to settle compensation for network participants, which can be used for interim settlement of balances stored in the network fee and database in DLT.

The first thing to consider when implementing the proposed platform is the basic operating protocol of the platform. The example platform operates a data use protocol based on the principle of least privilege. As a first step, it is possible to assume a scenario in which personal data is stored in an individual's smartphone and used locally. In fact, in most cases, that alone achieves the desired purpose. For example, in the case of target marketing based on hobby information, it was common for companies to market based on the hobby information itself in the form of raw data. Still, on this platform, hobby information is stored only in the terminal. A third-party app (e.g., Amazon shopping app) advertises based on the terminal's hobby information (e.g., golf). In this process, since personal hobby information is not stored in the server (e.g., Amazon server), consent to collect personal data is unnecessary. Some may ask if a third-party app stores it, but the individual also owns the app installed on the user terminal and its sandbox area as part of the individual's terminal, so it does not fall within the scope of information collection by the server. The next step is to use de-identified or anonymous data. Along with the limited authority provision, it is important to provide personal data only at the necessary level. It is important to de-identify personal data through various methods to minimize the risk of personal data leakage or exposure. Making tracking and accumulating personal data impossible through de-identification is also the most basic of protecting privacy. Data anonymity can be guaranteed when a technology such as blockchain is used.

The final step is to design it so that the usage history of the data can be tracked if the data itself is transmitted. Through a mechanism that digitizes the right to use personal data defined above, personal data can be kept from an individual's device, and its usage history is stored in a blockchain network as a ticket. Suppose personal data is managed on a personal smartphone, not a company's server. In that case, there is no need to sign in to the service individually to modify or destroy personal data. By setting the type and level of personal data available to companies in advance through the blockchain's smart contract, personal data can be freely viewed and/or used within the contract. In addition, the blockchain is open to network participants, ensuring the transparency of smart contracts for the use of personal data, drastically reducing the burden of proof of compliance for companies. In addition, the personal data used by inquiring directly to the individual is not the data provided when signing up but is accurate real-time personal data managed directly by the individual on the smartphone, which motivates companies to participate in the blockchain more.

Mobile Device-based MyData Platform/Eco-System Proposal

I will explain the proposed MyData platform by dividing it into technical and operational parts. Technically, based on a blockchain that guarantees anonymity, it is necessary to implement all personal data storage and processing in a form that only occurs on an individual's smartphone. De-identification processing is necessary because personal data has already been stored and processed based on identification. If there is no subscription process to join a separate ecosystem, anonymity is guaranteed. Just as no central authority is needed to create a Bitcoin wallet, anonymity is guaranteed when creating a personal data wallet for an individual; it can be solved if anonymity is guaranteed. All personal data is stored only on an individual's smartphone in the form of an on-device PDS described above, and the use of personal data can be controlled in the form of permission within the device. Based on this, designing a MyData platform by trading data access rights rather than data transactions on the blockchain is possible. All personal data is stored and managed only on smartphones. Still, it is implemented so individuals can directly control it by setting the type, period, number of times, compensation, and purpose (statistics, marketing, etc.) of personal data through the blockchain's smart contract.

Regarding operation, the aforementioned data access rights can be traded as tickets. We are all already familiar with this. We pay when we go to the theater and watch movies. When we watch Iron Man, Marvel makes money through the theater. Even if you watch Iron Man one more time or Marvel's Infinity War, you pay separately each time. Marvel is the content provider in this ecosystem. Marvel doesn't give us a movie film (raw data), but it sells us admission rights in the form of tickets, and we, as content consumers, are paying for it. It becomes a circular ecosystem only when there is a give-and-take. Similarly, for personal data, the individual is the content provider. Companies that are content consumers must pay for this to become a sustainable ecosystem. To this end, it is possible to replicate and reuse personal data indefinitely by distributing access rights in the form of tickets, not data itself. Although companies that are consumers of personal data do not receive personal data as it is, the data economy can be designed by selling and purchasing access rights from individuals who are providers of the content called personal data.

This design solves many obstacles we have considered a problem when utilizing personal data. The first is the issue of personal data processing. The idea of all kinds of privacy acts is that personal data does not belong to a third party. That's why there are complex consent mechanisms and obligations to destroy it. On the other hand, if an individual directly processes their data, there can be no compliance issue. The second is the assetization of data. Because data is intangible and can be replicated and reused indefinitely, it is valuable but difficult to become a quantitative asset. But it can be quantified if it is access rights, not data itself. Access rights can be quantified like Marvel's movie ticket, with access to my hobby information, access to it only once, and access to it for a year.

In fact, to some extent, these things are already familiar to us. Although not in the early days of smartphones, permission structures in Android and iOS have already introduced us to a mechanism for processing data within a device. Giving PERMISSION_CONTACTS permission in iOS does not mean that it collects the contacts of my acquaintances; it only uses the contact information within the smartphone to make calls and send text messages. Even if READ_EXTERNAL_STORAGE is granted in Android, it is not the right to collect all stored files but the right to access, open, modify, and store them within a smartphone. Expanding this to service data rather than specific data in the device can be a very flexible and powerful mechanism. These days, users' data is collected excessively, but ultimately, the only service they provide is to recommend customized credit cards, loans, and insurance products. No one recognizes the service as a MyData service but thinks it is an advertisement. For this purpose, rather than collecting personal data on the server, it is combined and stored only on the individual's device, and the recommendation algorithm itself can be executed on the individual's device. After that, companies can achieve their intended purpose by taking only the analysis results. In this process, companies can achieve their intended objectives without collecting personal data and even have the effect of ensuring real-time. If you think about it, whether I have $10,000 or $10,001 is not that important information for selling financial products. I must have more than ten grand in spare money. It doesn't matter if my birthday is in July or August. It is enough for companies to analyze and judge customers' interest in health-related products over their 40s. The problem is that collecting personal data is necessary because companies try to make such analyses and judgments. All these analyses and judgments are made on an individual's smartphone on the proposed platform.

Conclusion

The proposed MyData platform model aims to create a personal data compensation platform that benefits individuals and businesses. It has been presented as a sustainable personal data business ecosystem in which companies use personal data safely, and individuals are duly rewarded for personal data. Through the various MyData models that will emerge in the future, individuals should become true owners of personal data and be compensated for their use. Companies should be able to utilize personal data for business actively without legal risks. Due to the new personal data ecosystem implemented using blockchain, we expect a significant change in the paradigm of personal data use, which has never changed since the advent of the Internet.

In the future, personal data platforms and mechanisms will be attempted in various forms based on cloud or mobile devices. MyData will be attempted in the financial sector and various fields such as health care, e-commerce, administration, and education. Various attempts will be introduced for various innovative user experiences in the future. Still, humans must eventually be at the center of those attempts to achieve user experience and personal data protection.

Appendix

Government-led
MyData Case Studies in Korea

This chapter will cover case studies on MyData services in Korea. Although MyData was not introduced earlier than in other countries, the practical application is faster. In particular, MyData in the financial sector was quickly introduced led by the Financial Services Commission, followed by public MyData led by the Ministry of the Interior and Safety, and medical MyData by the Ministry of Health and Welfare is also in progress. While "MyData" is expected to be implemented in earnest from 2025, the medical field, including hospitals, pharmacies, and health information, is also included in the top 10 key sectors, expanding the scope and target of data transmission. The Personal Information Protection Act will be revised by the National MyData Innovation Promotion Strategy of the Personal Information Protection Committee and introduced in various fields of society in the future. So far, once the use of personal data is agreed upon according to the needs of companies and institutions, individuals will inevitably be passive in data utilization and management. In response, the Personal Information Committee firmly established that the people are the true owners of their data so that they can exercise their sovereignty over their data through MyData.

In particular, health care, telecommunication and Internet services, energy, transportation, education, employment labor, real estate, welfare, distribution, and leisure will be implemented first. The government plans to expand the data transmission and application scope gradually. The government operates a pan-government cooperation system to settle MyData in the entire field successfully. The pan-government MyData promotion team has implemented practical policies such as establishing a MyData legal system, promoting standardization, establishing and operating platforms, preparing infrastructure such as security and certification, and discovering leading services. In addition, a public-private joint MyData Council was launched under the supervision of the Personal Information Protection Commission, involving academia, industry, civic groups, and related ministries.

Financial MyData

Financial MyData was the first official MyData service introduced in Korea. The institutional basis of Financial MyData is the right to request the transmission of personal credit information (Article 33-2 of the Credit Information Act). MyData service refers to a service that integrates and manages the financial information of individual financial consumers. MyData service providers provide more convenient financial services to customers based on exercising the right to request the transmission of personal credit information, a right to strengthen the establishment of individual data sovereignty. For example, it provides an integrated inquiry service for personal account information by collecting financial information of individuals scattered in banks, cards, insurance, securities, and telecommunications in a lump sum and integrating it for customers to understand. In addition, you can present a list of financial products available under the individual's current credit and financial status and compare prices and benefits by-product in detail to receive financial product recommendations and financial consulting services optimized for individuals. The legal term for the financial MyData service provider is the 'personal credit information management business operator.' It must collect distributed personal credit information based on the exercise of transmission requests and provide services such as integrated inquiry, financial product advice, and asset management to customers. The Financial MyData business license process is divided into preliminary permission and main permission by the Financial Services Commission.

Personal credit information management business is divided into unique, concurrent, and incidental business. Unique business is to collect personal credit information held by credit information providers, users, or public institutions and to allow credit information entities to inquire and view all or part of the collected information. These include data analysis, consulting, information accounting, proxy exercise of information-related rights (e.g., the right to respond to profiling), training, education, publishing, advertising and promotion of financial products, and identification. According to the revised Credit Information Act, the personal credit information management business is operated under a permit system, and all companies that want to operate their own business must obtain permission from the Financial Services Commission. However, suppose the collected information is not personal credit information or does not provide the collected information to the data subject. In that case, it is not subject to permission, and the following cases are examples of businesses that do not correspond to the personal credit information management business.

First, if personal credit information is not processed, only personal information, not personal credit information, or only corporate credit information, not individuals, is processed. Second, it is a case where credit information providers and users (financial companies, etc.) or public institutions do not provide personal credit information. For example, a financial company's direct submission from a credit information entity (without exercising the right to request transmission of credit information) or providing only personal credit information generated directly during the financial transaction process is not financial data. Third, personal credit information is collected, but the collected information is not provided to the credit information subject by inquiry or reading. For example, financial companies collect personal credit information with individuals' consent to provide financial transactions, but this is the case in which they only use it for internal credit ratings without providing inquiry or reading to personal credit information subjects. Fourth, it is a simple household book application that cannot store and access personal credit information. Suppose personal credit information is not stored on a server of a commercial company such as a financial company, and it is impossible to access and inquire about the input personal credit information other than the data subject. In that case, it is also not subject to permission. Finally, cases without fear of hindering the protection of credit information subjects and sound credit orders are permitted under other laws and regulations and not subject to permission.

If necessary, the preliminary permission procedure may be omitted to expedite restructuring, customer protection, etc., such as merger or transfer of business, or meet the requirements for permission when applying for preliminary permission. The examination period for preliminary permission is two months, and the examination period for this permission is three months (one month for preliminary permission). The screening criteria include capital requirements and material requirements. It must have a capital of more than 500 million Korean Won and have all information processing and information and communication facilities determined and announced by the Financial Services Commission that it can properly process information. The feasibility requirements of the business plan shall be met. In particular, the business plan's organizational structure, management, and operation system shall be suitable for the promotion of the business plan. It shall not interfere with sound business, such as conflicts of interest and unfair practices. The screening criteria include capital requirements and material requirements. It must have a capital of more than 500 million won and have all information processing and information and communication facilities determined and announced by the Financial Services Commission that it can properly process information. The feasibility requirements of the business plan shall be met. In particular, the business plan's organizational structure, management, and operation system shall be suitable for the promotion of the business plan. It shall not interfere with sound business, such as conflicts of interest and unfair practices. In addition, requirements for the largest shareholder or major shareholder, requirements for the qualifications of executives, and requirements for the expertise of employment personnel will be reviewed.

As for my:D service of SNPLab Inc., it was interpreted that it is not subject to permission for financial MyData services in Korea. This is because when personal data is stored only in an individual's device, it is difficult to say that the platform operator literally 'processes' the personal data. And that was applied to combined data, too, because combining data is also a processing of personal data. Another combined MyData project funded by MOTIE (Ministry of Trade, Industry, and Energy) also concluded that it is not applicable for IRB (Institutional Review Board) obligation for medical data if an individual directly processes their medical data in their personal device.

Public MyData

Public MyData is a service that provides self-administrative information held by administrative and public institutions to the person or a third party at the data subject's request. For example, when the public applies for and receives services from administrative and public institutions or financial institutions, this includes a service that easily submits necessary document information at once. The legal basis is Article 10-2 (joint use of personal data at the request of the civil petitioner) and Article 43-2 of the Electronic Government Act (the right to request the provision of administrative information on the data subject).

Previously, it was inconvenient for the information subject to submit separate civil petition documents; prior consent was required when using administrative information jointly, and administrative information could not be inquired without the consent of the holding agency. The new service innovation is being applied to eight services currently operated by six institutions such as Small Enterprise and Market Service, Gyeonggi Job Foundation, Korea Credit Information Services, Credit Counseling & Recovery Service, Ministry of Health and Welfare, Korea Real Estate Board, associated with micro-enterprises, jobs, finance, etc., to enhance service provision to the people. Amid the recent growing importance of establishing data sovereignty in society, the Korean government has recently taken the following actions by introducing and promoting the 'Public MyData Service.' It has extracted the necessary data items from the documents for verification that citizens must submit to access various services offered by public and private institutions. It provided this extracted data in data packages, allowing citizens to utilize their administrative data directly. It has also streamlined procedures such as verifying document authenticity, reviewing documents, and inputting documents through the 'Public My Data Service' to help the institutions handling such tasks improve their job performance efficiency. MOIS has also created a 'Public MyData Portal,' a mobile service that can check administrative information available to the public and conveniently send administrative information to people wherever they want. There are three ways to send my administrative information from the Public MyData portal. 'Simple Sending' is a function that sends me my administrative information with one touch. 'Selective Sending' is a function that directly selects the desired item from the data items in my administrative information and sends it to the desired place (the people themselves, companies, or institutions). 'Regular Sending' is a function that allows you to send your administrative information to the desired place (the people themselves or companies or institutions) for a desired period (every week or month). Updated administrative information will be sent automatically during the period you set up. You can also check the public MyData service usage history (sent details, regular delivery details, and service usage details) at once on the public MyData portal.

The application process for utilizing public MyData is as follows. First, if you apply to the MOIS separately for each bundle data, you will receive a development tool (API/SDK) to receive data through the MOIS review. Subsequently, the receiving institution should develop the system using the

MOIS's development tool (API/SDK). To develop the system, the receiving institution must proceed with system development through the development tool (SDK/API) provided by the MOIS. Development is required accordingly since the received data standard (XML) is different for each bundle of information.

Public MyData is divided into five areas: health & medical, employment, start-up & management, life & safety, and welfare. HIRA (Health Insurance Review and Assessment Service) provides a service that allows legal representatives to automatically check their legal information only by mobile phone authentication without submitting evidential documents. However, it used to be necessary to check the history of administration of children under 14 years old. Gangwon province provides a service that allows you to apply online without submitting documents, unlike in the past, when you had to visit the city/county office to receive a worker deduction service that pays reserves to workers at maturity after accumulation every month. SEMAS (the Small Enterprise and Market Service) has simplified the document preparation process for applying for policy funds that small business owners must prepare annually through the platform. We supported the submission of 18 types of required documents that had to be submitted to the Small Business Development Corporation in a single batch of administrative information to apply for policy funds. The National Fire Agency allows 119 paramedics to quickly check their personal medical history and receive appropriate first aid and optimal treatment in emergencies. The Gyeonggi Job Foundation allows young people to apply for basic youth income services easily and quickly.

There are also cases in which local governments actively use MyData. SNPLab developed a service to prevent and respond to lonely deaths using artificial intelligence and MyData for Gyeonggi Province. Gyeonggi Province, the main provider of the service, began to provide "AI and MyData-based Solitary Death Prevention and Response Service" through the 2023 MyData Development Project hosted by the Ministry of Science and ICT and KDATA(Korea Data Agency).

Gyeonggi Province formed a consortium with Ansan City, KEPCO, SK Telecom, and SNPLab to demonstrate services for households at risk of dying alone in Ansan City. It predicts the risk of dying alone by fusing individual life data such as power, water supply, and communication usage status, and provides artificial intelligence care services through the "Gyeonggi Smart-D" platform.

It analyzes KEPCO's power usage data, SK Telecom's mobile phone usage data, and Ansan City's water supply use data for households at risk of dying alone. As a result of the analysis, notifications in the form of reports are provided to welfare managers and families according to the level of risk (normal to serious) and welfare managers at the administrative welfare center in the city of risk signs visit the site to check.

"Gyeonggi Smart-D" will be used to notify such risks and apply for welfare benefits. In addition, in Naver, artificial intelligence regularly calls households in need of care through 'CLOVA Care Call' to check their safety on topics such as health, meals, and sleep. If signs of crisis are found in the conversation or help is needed, monitoring contents are delivered to the administrative welfare center to prevent and manage them in advance. In a situation where the number of socially isolated households is increasing due to the increase in single-person households and aging population, the concept of MyData has been actively used to create a system that can respond more actively to the problem of lonely death. And to avoid the risk of excessive personal information inspection, SNPLab's on-device PDS technology and concept of federated learning were adopted.

Medical MyData

Medical MyData Service is a service that provides information on one's health to the desired target with one's consent so that it can be used in various fields, such as medical care and insurance. Medical MyData Service is helpful to the general public and medical service providers such as hospitals and clinics, healthcare-related companies, and research institutes. Medical MyData service is available to Koreans anywhere in Korea.

The Ministry of Health and Welfare in Korea is carrying out the following tasks to provide medical MyData service and is preparing a health information collection system for medical MyData service. Through this, health information held by individuals, medical institutions, and public institutions is collected and standardized according to each standard so that information can be transmitted smoothly. A health information highway platform has also been established to transmit health information effectively. Health information held by individuals, medical institutions, and public institutions may be effectively transmitted to places where necessary. Various measures are being prepared to develop related technologies so people can use health information independently. It supports technology to develop related technologies and systems so that health information transmitted through the health information highway platform can be effectively delivered to the desired place through apps with individual consent.

Healthcare MyData Services offers the following benefits to the public. The first is a customized healthcare service. Customized healthcare services that take into account individual physical characteristics will be possible. Personalized health care services (dietary, exercise methods, bio-rhythm checks, etc.) can be provided based on information that records individual physical characteristics such as individual heart rate, breathing, and muscle strength. Second, effective hospital care is possible. My health information is provided by sending the same information to any medical institution nationwide through the Medical MyData (Health Information Highway) platform with the individual's consent. Through this, no matter what medical institution you visit in the country, you can get the same health information and receive the diagnosis and treatment accordingly. Third, it provides various related institution service benefits. You can receive various benefits and information through the Medical MyData (Health Information Highway) platform with the consent of the government, local governments, public institutions, insurance companies, and pharmaceutical companies. Finally, it is expected to be used for research on advanced medical technology. I can donate or provide my health information to medical institutions, universities, research institutes, etc., to help develop new medical technologies such as vaccines and treatments, treatments for rare and incurable diseases, and medical devices.

Currently, 12 data items can be transmitted through the My Healthway platform. 'Patient data' represents personal data such as the patient's identifier, name, gender, and contact information as a resource for expressing patient data. 'Organization' is the identification symbol, contact, and address information of the medical institution, which are represented by resources representing the information of the medical institution. 'Practitioner' or 'PractitionerRole' indicates medical information and role information of medical treatment and is used as an essential value in the medical doctor of diagnostic details. 'Condition Profile' is information such as injury code, disease name information, and diagnosis date, expressed as resources representing diagnostic information. 'MedicationRequest' is a resource that expresses drug prescription history information and is used in the prescription date and time, prescription drugs, and how to take it. Observation represents information such as test code, test value, etc., as a resource for representing diagnostic test information. 'DiagnosticReport' represents information such as inspection code, inspection value, etc., as a resource that can represent inspection information. Another 'DiagnosticReport,' with the same resource name, includes information on DICOM (Digital Imaging and Communications in Medicine) images, such as the date and time of the shooting and DICOM series information. 'Observation' expresses reference information such as photographic equipment, photographic area, document information, etc. 'Procedure' is the resource for expressing surgery and treatment information, such as surgical code, date, and operation time. 'AllergyIntolerance' is an expression of information such as allergy cause code, and allergy code is expressed as a resource that can express energy and intolerance information. 'DocumentReferences' expresses document names, document types, etc., with resources that can express documents, such as proofs.

The Medical MyData platform consists of individual

health information, information from medical institutions, and related public institutions. These are collected and stored so that each institution can connect to and deliver the Medical MyData platform. These materials are used as basic materials to provide individual medical services. Personal health information collected and stored this way is transmitted through the Medical MyData platform with the individual's consent. The Medical MyData platform is connected to all medical institutions, research institutes, healthcare-related companies, and industries and can deliver information anywhere in the country. Health information sent with individual consent can be used for my health care and provided to medical institutions and research institutes such as hospitals and healthcare-related companies (bio companies, insurance companies, etc.) to help my health and life. In addition, considering that medical care is newly introduced along with the telecommunications and distribution sectors, a development strategy tailored to the characteristics of each sector is established, and data combination services between heterogeneous industries are expanded.

Leading medical services linked to Medical MyData will be used to exchange medical data, manage chronic health diseases, and manage medical prescription history by linking medical (test details, medical prescription information, health examination information), welfare (welfare supply information), and finance (insurance subscription details). It also introduced ways to link medical (chronic disease medical history), welfare (use of electricity, gas, water, etc.), and communication and Internet services (communication usage, smartphone wake-up frequency) to cope with emergencies in socially isolated households, such as preventing lonely death. In addition, in the medical field, it took as an example a development plan to strengthen industrial competitiveness by implementing patient care services without issuing unnecessary CDs and documents and creating a new medical data-based bio-health market. However, medical areas that require sufficient public protection, such as handling sensitive information on a large scale, will be operated under a permit system to designate specialized institutions separately.

About the Author

After graduating from the Department of Electrical Engineering at Seoul National University, Michael Jaeyoung Lee planned and established commercial services in the field of wired communication, using PON (Passive Optical Network) technology for the first time in Korea when working as a military service exemption in TriGem InfoComm (Subsidiary of TriGem Computer which was well known as eMachines). He also contributed to several professional technology magazines such as 'Management and Computer' and 'Network Times.' He later moved to the mobile industry to join TTPCom, a British company that provides GSM protocol stacks and mobile software platforms, to oversee mobile phone hardware and software development projects. After TTPCom was acquired by Motorola, he participated in the domestic feature phone and smartphone software development project as a lead researcher. After Google acquired Motorola, he participated in the device driver and OS upgrade project for Android smartphones.

At Samsung Electronics, he worked as a principal engineer, taking charge of software security assessment, security policy & privacy, and as a creative leader for C-Lab, an in-house venture program. In addition to several international public certificates related to security and privacy, such as internationally certified GDPR personal information protection experts (CIPP/E), internationally certified security experts (CISSP), and internationally certified ethical hackers (CEH), he has various qualifications, such as financial planners (AFPK®). He is interested in changing the paradigm of personal data use mechanisms. At (ISC)² The Security Society, he drew attention with his announcement of a new topic for personal data protection under the theme of "New Approach for Personal Data Protection," and designed and proposed an on-device MyData platform that operates on mobile terminals for the first time so that companies can safely utilize personal data.

He founded SNPLab Inc., a MyData platform company that provides comprehensive solutions for personal data utilization and has won various awards from the Minister of Science and ICT, the Minister of SMEs & Startups, the Korean Intellectual Property Office and the MyData Operator Award 2023 from MyData Global Association and is building a new type of MyData ecosystem based on on-device and edge computing technologies. He is also the author of the 'Fintech and GDPR' series, which were bestsellers for ten weeks in the Internet business sector in Korea, 'Data Sovereignty and

MyData,' and 'Hello, Fintech!' which is used for the fintech textbook of Korea Banking Institute by Financial Services Commission in Korea. He still writes many columns for magazines in various fields, including law, finance, and technology, and is appearing in many e-learning and seminars to improve individual digital literacy. He studied blockchain-based personal data platforms at Hanyang University's Department of Blockchain Convergence and teaches AI Technology Management at Kyunghee University.

References

Article 29 Data Protection Working Party (2017, October 13), *Guidelines on Data Protection Impact Assessment (DPIA) and determining whether the processing is "likely to result in a high risk" for the purposes of Regulation 2016/679* [European Commission Guidelines], https://ec.europa.eu/newsroom/article29/items/611236

Kim, Jaeyoung (2021). *A study on the consumer protection plan for MyData service*, Korean Consumer Agency, 74

KOREA Data Agency (2023). MyData Guidebook, https://kdata.or.kr/mydata/www/board/guide_04/boardList.do

Korea Health Information Service (2021). *Trends and implications of domestic and international laws and systems on data ownership, 2021 3rd Forum on Healthcare Data Innovation,* https://www.k-his.or.kr/boardDownload.es?bid=0025&list_no=357&seq=1

Lee, Dongjin (2020). *Legal issues and guidelines for data transactions*, SNU AI Policy Initiative, 11-13.

Lee, Dongjin (2018). *The Concept of Data Ownership—A Critical Observation*, Information Law, 219-242

Lee, Jaeyoung (2020). *Fintech and GDPR1*, Privacy protection for Fintech, Korea: Tarukus.

Lee, Jaeyoung (2020). *Fintech and GDPR2*, Approach for data sovereignty, Korea: Tarukus.

Lee, Jaeyoung (2021). *Hello, Fintech!*, Communication Books, Korea: Fintech Center Korea.

Lee, Jaeyoung (2021). *Data Sovereignty and MyData*, Korea: Communication Books.

Lee, Jongju (2019, October 15). *Data Ownership Trends*, Software Policy & Research Institute, https://spri.kr/posts/view/22802

MyData Global Association website and white papers, https://www.mydata.org/

Ministry of the Interior and Safety, (2022, Jun 21). *Publication of public MyData service usage guides and business guidelines* [Press release], https://www.mois.go.kr/frt/bbs/type010/commonSelectBoardArticle.do?bbsId=BBSMSTR_000000000008&nttId=92639

National Research Foundation of Korea. (2022, July 4). *The Impact and Implications of Blockchain Technology on MyData and Artificial Intelligence Ecosystems in the Web3.0 Era* [NRF Issue Report], 2022-7, https://www.nrf.re.kr/file/boardDownload?fileType=N&no=180197&sno=1&menu_no=419

National Research Foundation of Korea. (2021, September 13). *Fintech And MyData In Post-Covid Era* [NRF Issue Report], 2021-17, https://www.nrf.re.kr/file/boardDownload?fileType=N&no=163381&sno=1&menu_no=419

Polonetsky, J., & Renieris, E. (2020). *10 Privacy risks and 10 privacy enhancing technologies to watch in the next decade*, Future of Privacy Forum, 7-9.

Public MyData Portal (2022, Jun 20). Public MyData Service Agency Performance Guide, https://adm.mydata.go.kr/cmm/fileDownload3.do?fileSeqno=397

Personal Information Protection Commission in Korea. (2023, August 3). *Policy Directions for Safe Use of Personal Information in the Age of Artificial Intelligence* [Press release], https://www.pipc.go.kr/np/cop/bbs/selectBoardArticle.do?bbsId=BS074&mCode=C020010000&nttId=9083

Reinsel, D., Gantz, J., Rydning, J. (2017). *Data Age 2025: The Evolution of Data to Life-Critical*, An IDC White Paper , Sponsored by Seagate, https://www.seagate.com/files/www-content/our-story/trends/files/ Seagate-WP-DataAge2025-March-2017.pdf

Smart Health Standards Forum (2021, November 30), *Final Report on the Preparation of the Korean Technology Standard for FHIR (Fast Healthcare Interoperability Resources)*, https://k-his.or.kr/boardDownload.es?bid=0015&list_no=826&seq=1